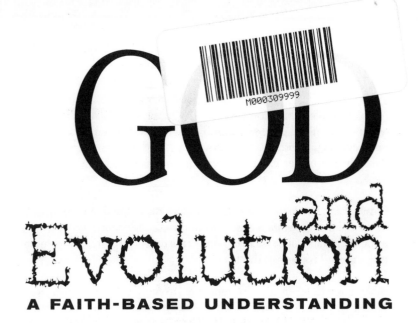

GOD and Evolution

A FAITH-BASED UNDERSTANDING

DAVID L. WILCOX

JUDSON PRESS

VALLEY FORGE, PENNSYLVANIA

God and Evolution: A Faith-Based Understanding

Judson Press has made every effort to trace the ownership of all quotes. In the event of a question arising from the use of a quote, we regret any error made and will be pleased to make the necessary correction in future printings and editions of this book.

Bible quotations in this volume are from *The Holy Bible*, King James Version. (KJV) and HOLY BIBLE: *New International Version*, copyright © 1973, 1978, 1984. Used by permission of Zondervan Bible Publishers. (NIV)

Library of Congress Cataloging-in-Publication Data

Wilcox, David L., 1943–
    God and evolution : a faith-based understanding / David L. Wilcox.—
1st ed.
        p. cm.
    ISBN 0-8170-1474-8 (pbk. : alk. paper)
    1. Evolution—Religious aspects—Christianity. 2. Bible and evolution.
    3. Religion and science. I. Title.
        BT712.W54 2004
        231.7'652—dc22

                                                                2004017028

Printed in the U.S.A.

10 09 08 07 06 05 04

10 9 8 7 6 5 4 3 2 1

To *my father,*
Dr. D. *Ray Wilcox,*
who put me on this track
nigh 50 years ago
by pointing me in the direction
of finding my own answers
rather than telling me
what I had to believe.

CONTENTS

Introduction

" **I** CAN'T BELIEVE IN BOTH GOD AND DINOSAURS. SO I PICKED God." These words from a nine-year-old girl represent the sentiments of thousands of Christian people who, based on the perception that dinosaurs are not mentioned in the Bible, feel compelled to choose between their faith and dinosaurs. Many also feel compelled to choose between God and evolution, as they understand it. This choice faces third-grade girls and eighty-year-old grandparents, parents and teens, pastors and biology students.

The subject often is a source of tension and confusion. In colleges and universities across the nation, including Christian institutions, what students encounter in biology class with regard to the ideas, theories, and facts associated with evolution is very different from what they learned in Sunday school. The experience can be disorienting. Many are convinced that science is attempting to dethrone the Creator God and install Nature on God's throne. Some are ready to fight, and others to run.

For over thirty years I have observed these reactions. I have spoken with students (and with many of their parents) about the implications for faith of various scientific facts, theories, possibilities, and probabilities. And I have wrestled personally with the relationship between my Christian faith and the data gathered from proven, reliable scientific methods. The good news is that I am convinced that there is a point of resolution. We do not need to choose between God and dinosaurs, between faith and evolution (depending in part, of course, on how evolution is defined and understood). There *is* a way for God to be both our God—the God of the Bible and the God of Christian faith—and the God of cave dwellers and dinosaurs. Part of the purpose of this book is to present and defend this position—to suggest new ways of looking at old things, to offer a perspective of faith that does not ignore facts or compromise scientific integrity.

I invite you to come with me on this challenging and interesting journey. Together, we will examine the group of ideas collectively referred to as "evolution." We will explore various definitions of evolution and attempt to determine which are compatible with a Christian worldview and which are not. We will critique the science that supports various aspects of evolutionary theory. We will ask: Is this thing called "evolution" an instrument of the devil? Or is it simply neutral science? Is it perhaps a description of God's work in creation? Or might it be (again, based on how we define the term) a mixture of all these three?

The debate between "creationists" and "evolutionists" typically centers on questions such as "Is the earth a few thousand years old or a few billion?" and "Are apes the biological ancestors of human beings?" In the later chapters of this book we will consider these and other important questions.

No doubt, some of the material in this book will seem too basic for some and perhaps a bit too complicated for others (though I have tried to keep it as simple as possible). Such are the hazards of attempting to reach an audience that spans several different levels of familiarity with biological terms and principles. When the scientific jargon is complex and technical, I will do my best to make clear the implications, the significance, of the evidence.

It behooves us, however, before rushing too quickly to the center of the debate, to explore some foundational questions, the answers to which will serve us well in terms of putting the debate into its proper perspective. With this in mind, in the early chapters we will explore topics such as a biblical theology of nature and how science works. We will examine the history of the conflict between science and theology and discuss the proper domain of each, in part with an eye toward clarifying the terms of the debate and adopting ground rules by which opponents might disagree without demonizing or misrepresenting those who hold different perspectives.

As Christians, we have an opportunity to explore the marvelous world that God has created and sustains. This includes exploring the history of that world and the history of God's crowning achievement: human beings. We explore with a sense of adventure, confident that because all truth is God's truth, we need have no fear of what we might discover.

CHAPTER 1

What the Bible Says about Nature

NO MATTER WHAT THE TOPIC, THE STARTING POINT FOR Christians is the Bible. And so we begin by examining what Scripture has to say about nature, about the natural world. The natural world is, after all, the object of scientific study. Thus it is a central aspect of any discussion about evolution.

Simply put, the essence of the biblical message is that God made the natural world, and God governs it. Let us consider a few verses that address how God acts in and upon the natural order. Read straight through the following Bible verses in order to get a quick impression of how the Bible's writers saw God's creation.

"He [Christ] is the image of the invisible God, the firstborn over all creation. For by him all things were created: things in heaven and things on earth …; all things were created by him and for him. He is before all things, and in him all things hold together" (Colossians 1:15-17).

"He commanded and they were created. He set them in place for ever and ever; he gave a decree that will never pass away" (Psalm 148:5-6).

"He has spoken to us by his Son, whom he appointed heir of all things, and through whom he made the universe. The Son is the radiance of God's glory and the exact representation of his being, sustaining all things by his powerful word" (Hebrews 1:2-3).

"We were chosen ... according to the plan of him who works out everything in conformity to the purpose of his will" (Eph. 1:11).

"Not one of them [sparrows] will fall to the ground apart from the will of your Father" (Matthew 10:29).

"These all look to you to give them their food at the proper time. When you give it to them, they gather it up; ... when you take away their breath, they die and return to the dust. When you send your Spirit, they are created, and you renew the face of the earth" (Psalm 104:27-30).

From these passages (and there are many others) we can begin to draw conclusions. First, God did not "begin." Nothing made God. God exists in eternity. After all, something must exist eternally—something ultimate before everything else, something timeless, something that makes everything else happen. For the Christian, that something, or more accurately, that *someone,* is God.

God made everything because God wished to do so. God set nature (the creation) in order, established its laws, assigned to the various parts of nature their tasks and directed them toward his goals. Nature was created to be obedient as God's servant. Its purpose was to display God's glory, and many, especially those who have studied the world from a Christian perspective, would agree that nature has been very good at fulfilling this purpose.

According to Scripture, God, after creating, did not just go away. Rather, God has continued to be in an active relationship with nature. Nature, lawful and obedient, follows the order set up in creation, while at the same time God continues to fully govern events. God remains aware of and determines the mundane details of nature, weaving them all into the tapestry of his eternal purposes. Simply put, the God of the Bible is the Shepherd-King of nature.

In answering prayers, God need not break the created flow of natural cause and effect. Nor does God "wait" to answer, for God remains outside the flow of time. God can shape the whole space-time continuum, answering a child's prayer without changing the orbit of Jupiter or anything else. It is true if God wants to catch our attention in a special way, God may temporarily amend a law or two and change his ordinary way of governing nature. But even such a special event (sometimes called a "miracle" or a "singularity") is not an afterthought. Rather, God uses these events to show evidence of the divine presence, not to "fake us out" when we are trying to understand nature. Nor does God deviate from his usual methods because he became stuck and had to do things a certain way.

The Christian view of nature that I have just described has been, consistently throughout church history, the church's understanding of the relationship between God and nature. The French Reformer John Calvin wrote in his *Institutes of the Christian Religion:*

> Moreover, to make God a momentary Creator, who once for all finished His work, would be cold and barren, and we must differ from profane men especially in that we see the presence of divine power shining as much in the continuing state of the universe, as in its inception....Faith sought to penetrate more deeply, namely, having found him Creator of all, forthwith to conclude He is also everlasting Governor and Preserver—not only in that He drives the celestial frame as well as its several parts by a universal motion, but also in that he sustains, nourishes and cares for everything he has made.[1]

In chapter 5 of the Westminster Confession of Faith (the creed of the Scottish Presbyterians and English Puritans), on the subject of providence, the writers stated three articles:

> **1.** God the great Creator of all things doth uphold, direct, dispose, and govern all creatures, actions, and things, from the greatest even to the least, by His most wise and holy providence,

according to His infallible foreknowledge, and the free and immutable counsel of His own will, to the praise of the glory of His wisdom, power, justice, goodness, and mercy.

2. Although, in relation to the foreknowledge and decree of God, the first Cause, all things come to pass immutably and infallibly: yet, by the same providence, He ordereth them to fall out, according to the nature of second causes, either necessarily, freely, or contingently.

3. God in His ordinary providence maketh use of means, yet is free to work without, above, and against them, at His pleasure.[2]

God is the cause of all that happens, everything happens according to God's plans, and most of the time God brings these plans about through "second causes" (i.e., natural causes, cause and effect in nature), although God does not have to do it that way. As C. S. Lewis put it, God is free, "not a tame lion."[3]

Thus, whatever we see happening in nature occurs because God is directing it to happen according to his own laws. People may disobey, but nature obeys. In nature we do not see *everything* that God is doing, but neither do we so see *anything* that God is not doing. Nature is not running "on its own." It is neither a programmed clock nor an out-of-control juggernaut. Rather, natural processes testify to God's directing hand (providence).

When the Scriptures convey that God is the Lord of nature, they provide both the justification and the confidence to study nature. As we engage in a scientific study of the world that God has created and sustains, we can trust that the laws of nature will be consistent and reliable, that they will tell us the truth. We can trust that God does not mess up in his attempt to allow the natural world to testify to his glory. Thus, a Christian theology of nature provides a perfect foundation, a solid rationale, for believers to be scientists pursuing an understanding of all God's creation. It is with this understanding that we now move forward.

CHAPTER 2

Understanding Science
and Its Limitations

A S WAS SUGGESTED IN THE FIRST CHAPTER, PEOPLE PRACTICE science in order to figure out how nature works. This statement, though short and simple, contains three very important concepts about science: (1) it is *people* who do it; (2) it involves the *mind* ("figure out"); (3) it is focused on *nature*.

The Scientific Method
Let us start by describing, in a somewhat idealized and basic way, what happens in a scientific investigation. An investigation begins with an investigator, a person, who observes something that attracts his or her attention as being significant, something worth exploring and perhaps explaining. At some point the investigator notices what appears to be a pattern of events. Perhaps the investigator has a feeling that a majority of traffic accidents involve red cars. Or perhaps the investigator has observed that HIV patients seem to lose the white blood cells known as T4 lymphocytes. In both examples the observed relationship between one thing and the other can be termed a "possible correlation."

A trained investigator should know that the mind sometimes plays tricks. Thus the first step is to document the pattern—that is,

5

to determine if what appears to be a pattern truly is one. After all, we have learned many patterns (memories, templates in our brains) in our lives, and we use this information to recognize other patterns. But we know that sometimes the learned patterns mislead us. And sometimes we assume that we see patterns, but they are not really there.[1]

So how do we check our intuitions? The answer is, in a word, statistics. The whole purpose of statistical testing is to help us separate real patterns from false or misperceived ones. A "95 percent confidence limit" means that a particular pattern holds true to form nineteen times out of twenty, which is plenty often enough to consider it a genuine pattern, though in some cases it might make sense to ask why a pattern does not occur 100 percent of the time. If a pattern is in doubt, we may need to collect more observations (termed "data") to make the statistical test results more certain, one way or the other.

Once an investigator is sure that a pattern is real, the next step is to try to explain it. In science we term potential explanations of patterns in nature "hypotheses" or "theories." They constitute a response to questions such as these: How do HIV patients lose T4 lymphocytes (white blood cells)? Why do drivers of red cars have more accidents? How did a gambler flip a coin and get ten heads in a row?

Obviously, a proposed explanation is not always a correct one. We could theorize that the gambler has a coin with heads on both sides, but that might not be so. We could theorize that a high percentage of careless drivers buy red cars, or that red cars somehow make drivers careless. In the case of the gambler, we need only examine the coin to test the double-heads theory. Testing either of the red car theories is not so simple.

How do we test a theory? When it comes to investigating nature, the scientist in essence goes to nature and says, "Here is what I think. Am I right?" The scientist asks this question of nature through prediction and experiment. If a particular explanation (theory) is correct, then the predicted results should follow specific actions. For instance, if it is true that careless drivers tend to buy red cars, then one would predict that their driving records prior to the red-car purchase will reveal a higher accident rate. The process set up to collect such information is called the "experimental design." We test the theory by collecting the data, and the data will tell us whether we were right or wrong.

So what happens next? What if through experimentation nature apparently tells us that a particular pattern is real? The traditional understanding of the next step in the scientific process is to group a proved theory with other patterns in order to understand and explain larger patterns or principles in nature. These larger patterns sometimes are called "paradigms." A "good" paradigm is one that proves useful, one that suggests new theories that in experiments consistently produce expected results.

But does that take care of it? Is that enough? Is finding truth as simple as that? Not on your life! Imagine a police investigation in regard to a missing child. The detectives are on the job. They have some ideas and a few suspects, but so far the conclusion that they have reached does not allow for much hope. In contrast, the family clings to every positive lead, hoping against hope. They anticipate a different result. Sometimes the results we expect are influenced by a bias of one kind or another. We all have biases, paradigms that help us make sense of the world, even if they do not help solve crime.

In a murder investigation, as in science, the critical distinction is the one between data and theories. Theories may be wrong, but the data can never be wrong or right. They simply are what they are. Rifle cartridges, blood drops, clothing fibers, and fingerprints are data—what a detective calls "evidence." No one can go back in time and watch the crime. The explanations for who might have fired the gun, lost the blood, or touched the car handle are theories. The world-weary expectation (the paradigm) of the cops that the jilted boyfriend is the likely suspect will lead them to theories linking him to the disappearance. The desire of the family for the girl's safe return will lead them to theories with a more hopeful outlook. Detectives will search for more data, looking in places suggested by their theories, hoping (or perhaps fearing) to confirm their suspicions.

We must realize that data are not theories; they do not come with attached explanations as to what they mean. However, additional data might lead to the modification of a theory. And theories in turn might aid us in recognizing and organizing data. Of course, not all of what is considered data is necessarily valid when it comes to proving a particular theory. Wishful thinking can act in science as easily as it does in police work. But genuine data cannot be wished away when it does not fit into our accepted view of reality. If on the next morning the missing girl calls home from a friend's house, clearly she is alive. No homicide detective's theory can change that.

Proving theories from data is not as simple as it may seem to some. A validated prediction, for example, does not necessarily prove the theory. There may be an alternative explanation for the data, one that results in the same fulfilled prediction.

Sometimes the predictions that we make are too closely tied to our background assumptions or "shaping principles." When we do not get the predicted results, we simply revise our assumptions rather than drop the theory. Those background assumptions, our biases, are the necessary framework that we use to make sense of reality. They are our paradigms, how we make sense of our world. As such, they sometimes are hard to abandon. And so instead of giving them up, we bend and stretch them in order to allow for the data to fit somehow.

Certainly paradigms play a critical role in doing science. They are the spectacles through which we look at reality. Data do not produce their own theories. Rather, paradigms produce theories resulting from data patterns. Again, a good paradigm is one that proves useful as an investigator thinks of theories that he or she can test experimentally and get positive results. Investigators tend to hang on to paradigms because they represent the larger ideas that make sense of the smaller ones. But when a paradigm user is faced with patterns of data that "do not make sense"—that is, when a paradigm loses its ability to explain things, when the paradigm no longer can produce good, testable theories—then it may be time to change the glasses, to adopt a new paradigm and see the world in a different way. This we term a "paradigm shift" or a "scientific revolution."[2] A historical example of such a shift is the paradigm of Copernicus that the earth orbits the sun, which replaced the Ptolemaic paradigm that the sun orbits the earth.

The Limitations of Science

Some people tend to think of science and the scientific process as being fully objective, precise, foolproof. But science is far from infallible, especially when it comes to interpreting data. After all, science is done by human beings, who ultimately are limited in their understanding, extensive though it may be, and who are far from infallible. Sometimes, particularly when we are dealing with complex problems, the efficacy of experimental designs is open to debate. Or the design might be flawed in ways that the investigator cannot see. Sometimes investigators hang on to paradigms that have been useful in the past, even if new information and understanding cast doubt on their accuracy.

We tend to think of the scientific data—"the cold, hard facts"—as being completely objective, and in many cases they are. That is, some scientific facts are undeniable, or virtually so. But in many instances interpreting the data, which includes determining which objective facts are most relevant in explaining a particular phenomenon, is a highly subjective endeavor about which intelligent and fair-minded people may disagree.

9

Beyond all this there are questions related to value, purpose, and morality that science simply is not qualified to address. This is the domain of Scripture and of theologians and philosophers. For example, science can conclude that without treatment, tuberculosis bacteria in your lungs will multiply and kill you. But science cannot tell you whether it would be better for you to live than to die. It cannot determine for you whether your death from tuberculosis is worse than the death of the tuberculosis bacteria from the antibiotics. Science cannot tell you whether a doctor's choice to withhold your treatment is good or evil, because these are value judgments.

Even if "science" could explain in some way the source of your feelings of morality—for example, if it could explain why you consider it wrong to toss out the tuberculosis antibiotic—ultimately it would not matter, because even this explanation would not address the question of the validity of those moral feelings. Questions related to morality go beyond the mere consideration of feelings. Is a murder acceptable if the murderer has no feelings of guilt? Even if all people in the world were to agree that murder is morally acceptable, does that make it so? Again, science is incapable of answering such questions, questions that lie outside the boundaries of a physical study of nature. Thus science cannot issue judgments on values, ideas, or principles that are rooted in religious faith.[3]

Much of the tension in the debate over evolution results from flawed understandings regarding the boundaries of both science and theology. Conflict arises when these boundaries are not respected. It is to this topic that we now turn our attention.

CHAPTER 3

Science Versus Theology

A S WE HAVE SEEN, THERE ARE SOME THINGS THAT SCIENCE CAN-
not explain. And there also things—issues of morality and
values—that science, defined as the study of nature, does
not even address (even though some scientists are prone to cross
this boundary). Yet in one sense it would be inaccurate to say that
science and faith are totally separate. In fact, one could argue that
a kind of faith principle underlies the very process of science, thus
creating the possibility for the study of nature in the first place.

Remember that science is the human effort to understand nature.
But for people even to expect to be able to do that, they must make
some assumptions. A scientist might claim to start with no assumptions, but as we will soon see, that is impossible. First, when we
investigate nature, we assume that indeed there is a true and objective reality that exists outside of our own perceptions, preferences,
or beliefs. We posit (or assume), for instance, that we are not insane
and that reality is not merely a product of our own minds.
Furthermore, science assumes that this objective reality could have
been other than what it is, that its nature is contingent on some
determining factors. (Although both Christianity and science make
these assumptions, many human cultures do not. For instance,
some major Eastern religions, such as Buddhism, consider physical
reality an illusion.)

Second, in order to undertake any meaningful or relevant study of nature, we must assume that the universe (nature) has a rational order. That is, we assume that natural law makes sense, that specific causes produce specific effects, even if we cannot always understand them completely. If we thought that the universe was a disorderly chaos, void of any predictable patterns, then we would not bother to try to understand it. Without cause and effect in nature, the very concept of science as we know it would be meaningless.

In searching for order in nature, science must first assume that such order exists. This assumption, known as the "uniformity principle," includes the assertion that natural laws do not change. This assumption is necessary for science, even though it cannot be scientifically proven, for any and all proofs would have to begin with the assumption.

Third, as scientists we must assume that human beings are capable of understanding nature. This means assuming that we are rational and that our senses can be trusted. If ultimately our minds are not rational, then we are incapable of understanding the true reality of what we perceive. The ideas that come to us may enable us to survive, but this does not mean that they are firmly tied to ultimate reality. To do science, we must assume this connection. In short, we cannot prove that we can prove anything. So some things are assumed.

Now if we are going to be comfortable with these assumptions, they need to make sense in the context of our overall view of reality. What sort of a world would have an understandable, contingent, rational order? What sort of a reality can assure us of our own rationality? The answer is that if the source of all things is itself thought to be rational, then all the preceding assumptions make sense. If the source of all things is thought to be chaotic and mechanistic, then these necessary assumptions must be taken on blind faith. Thus, in theory, Christianity is fully compatible with science, for according to Christianity, the ultimate reality that shaped an obedient nature and the thinking human mind is itself rational. This ultimate reality is God.

Examining the Conflict

To be sure, it is easy enough simply to say that science goes to nature for truth, while Christianity goes to the Bible. But what if nature and the Bible appear to be in conflict? In that case, how can both be reliable sources of truth? How can both be right?

As we have seen, science assumes nature to be in some sense authoritative. Nature is what it is, no matter what an investigator thinks or wants it to be. Technically speaking, nature can never be in error. Nature cannot be right or wrong. It simply is. Furthermore, for the Christian, nature is as it is because God decided that it should be just that way. Thus, nature's authority is God-given.

However, though nature cannot be in error, science (and scientists) certainly can be in error, for science is not nature itself, but rather is our human understanding of nature. Science is our human interpretation of nature, and our potentially flawed understanding and interpretation of nature do not constitute absolute truth.

Does this mean, since science is subject to error, that theology is "queen of the sciences"? Absolutely not! To be sure, if the Scriptures are "God-breathed" (see 2 Timothy 3:16)—the truth of God spoken through human beings led by the Holy Spirit—then the Scriptures cannot be in error. Thus, the Bible's authority, as is the case with nature, is God-given. But although the Bible, like nature, cannot be in error, theology can indeed be in error. Just as human scientists can misunderstand nature, so human theologians can misunderstand and misinterpret Scripture. Theology is not Scripture itself. Like science, it is not absolute truth. It is prone to error.

Both science and theology have the same source of authority: God. And both fields of inquiry have the same faulty interpreters: human beings. If the Bible is from God and nature is from God, then the Bible and nature cannot be in conflict. Thus any conflict that arises represents flawed human understanding of nature, of Scripture, or of both. Not surprisingly, both theology and science tend to blame the other for the errors that give rise to conflict.

Later on in this book we will consider some guidelines for resolving the conflict between scientists and theologians. Early on, however, it is important to call attention to one very common and important aspect of interpretation. Whether interpreting the Bible or nature, both theologians and scientists are well advised not to accept what they encounter at face value—that is, the "literal" meaning. By "literal" I mean the interpretation that first comes to mind upon reading the text of Scripture or observing the phenomena of nature. In some cases the scientist or theologian might end up accepting the literal (or seemingly obvious) meaning. But one should not allow first impressions, however strong or seemingly clear, to limit further exploration. Keep in mind that observations can be a lot like Rorschach inkblot tests, which are designed to tell us more about ourselves than about the inkblot. Likewise, what we see or hastily conclude in a situation—human, natural, or scriptural—sometimes tells us more about our own mental patterns or preformed judgments than it does about whatever reality we are observing and trying to understand.

When it comes to interpreting the Bible, we must be aware that some Scripture passages were never intended to be interpreted "literally" by our twenty-first-century minds. For example, most people have heard of "the twelve tribes of Israel." The Bible refers to the tribes frequently, often listing them. The truth is that there were actually thirteen tribes. One of Jacob's twelve sons, Joseph, had two sons, Ephraim and Manasseh, each of whom gave rise to a tribe. However, when the tribes are listed in the Bible, one (usually either Levi or Benjamin) is always left out. Does this mean that the Bible is in error? Based on a "literal" reading from the vantage point of our twenty-first-century mathematical culture, some would say yes, because for us numbers mean only "how many."

To the Israelites, however, numbers also connoted qualities. Specifically, "twelve" connoted "the people of God." Thus the intended meaning is that the Israelites were God's chosen people. To have said "the thirteen tribes of Israel," though more mathematically accurate to our modern-day minds, would have worked against the primary truth that the biblical authors were trying to convey. Whether, technically, there were twelve or thirteen was not the point. Thus we see that it is possible for the perceived "literal" or face-value meaning of Scripture to shield us from understanding its true theological meaning.

A Brief History of the Conflict

Despite the widespread perception that the conflict between science and Christianity is a mostly contemporary phenomenon, the history of this conflict dates to ancient times. The conflict is actually not between science and Christianity, but between two philosophies—two different ways of making sense of the world, two different sets of basic assumptions. One such philosophy is materialism, the idea that matter is all that exists and that matter orders itself unguided and without purpose. The second, competing philosophy is supernaturalism, the idea that some reality exists outside of matter, and that this "extramaterial" (spiritual) reality provides purpose and order to the material world. Christianity is, of course, a form of "extramaterialism" that falls under the category of theism. In Christian theism the outside reality, the God of the Bible, is considered to have rationality and personality. As for science, it is not correctly labeled as either a materialism or as an "extramaterialism." That is, because science is the study of nature, it can be conducted under either a materialistic or an extramaterialistic set of assumptions.

The argument between materialism and theism has been going on for at least 2,500 years. The Greek philosopher Democritus argued for materialism, while Plato argued for a form of theism.

The Roman writer Lucretius (an Epicurean) argued for materialism; Cicero (a Stoic) defended the gods.

Under the dominion of the medieval church, the materialist position mostly vanished in favor of Christian variants of Plato's views. Debates continued, of course, including debates over how nature was to be studied. The "rediscovery" of Aristotle in the thirteenth century spawned a great deal of this debate, which for the most part remained within the church (i.e., within the universities).

As the cultural power of the church began to break down in the sixteenth century, the anticlerical movement known as the Enlightenment became increasingly materialistic. Modern science was underway at the time, but this "natural" science was being conducted almost entirely by investigators (including Robert Boyle and Isaac Newton) who subscribed to the theistic position.[1] However, whenever possible, Enlightenment thinkers such as Voltaire used the changing scientific views of nature to attack both the church and the validity of the Scriptures. Theists, on the other hand, used scientific discoveries to defend their positions (an effort called "natural theology").

Historian James Moore documents that around 1840 a new movement of young middle-class reformers appeared in England, calling themselves "naturalists."[2] As young adults, they had typically changed their creed from Christianity (which they felt was morally bankrupt) to a creed based on "nature." (This is not a "scientific movement." The group consisted of "poets and lawyers, doctors and manufacturers, novelists and naturalists, engineers and politicians." Notice the professions of nineteenth-century naturalist luminaries such as novelist George Eliot, poets Alfred Lord Tennyson, A. H. Clough, and Matthew Arnold, biographer J. A. Froude, industrialist Charles Bray, theologian F. W. Newman, social reformer Harriet Martineau, sociologist Herbert Spencer, mathematician Francis Galton, neurophysiologist G. H. Lewes, physicist John Tyndall, and, of course, "Darwin's bulldog," T. H. Huxley.)

The new creed of the naturalists redefined human nature, society, order, law, evil, progress, purpose, authority, and nature itself in terms of their particular view of nature (a new version of materialism) rather than in terms of Scripture (which they said was the true source of society's evil). God, if God existed at all, could be known only through nature.

Naturalism attempted to take control of the culture away from the established church. For instance, T. H. Huxley's "X" club engaged in a successful campaign to take the university chairs in the sciences away from the Anglican clergymen/amateur scientists. Now remember that naturalism is a social movement, not science. However, if Nature replaces Scripture as the primary (or sole) source of truth about morals and purpose, then the scientist replaces the priest, and the lectern becomes the pulpit.

The theists did not go quietly into the night, of course. Christian scientists and theologians alike argued that the assumptions and conclusions of naturalism were both untrue and unnecessary. They showed how the new (nineteenth-century) science (including many of the ideas associated with evolution) was still compatible with Scripture. And in fact, theists in science are still with us today: A 1997 survey showed that 40 percent of scientists still believe in "a God who answers prayer."[3]

Nevertheless, the naturalists for the most part succeeded.[4] They took over the universities. Huxley's "Young Guard" used the trappings of religion to sacralize their "science." Three centuries of cooperation between science and religion were forgotten, and their history was rewritten as "warfare." Hymns to nature were sung at popular lectures before the presenting of "lay sermons" by a member of Galton's "Scientific Priesthood." Museums were built to resemble cathedrals, and Darwin was buried in Westminster Abbey. The new "church" was firmly established.

Today, of course, the "church" of naturalism is alive and well. Its continued success can be judged by the general acceptance given the pronouncements of the "true believers" of the "church scientific." In modern times God has continuously been publicly (and negatively) discussed by some very well-known scientists, including Richard Dawkins, Daniel Dennett, Stephen Gould, Carl Sagan, Edward Wilson, and others.[5] It should be understood that when such persons bring God into the discussion, they are by no means speaking as scientists, even though some may think that they are doing so. Rather, they are using up-to-date science to engage in worldview apologetics, defending a view according to which nature functions as God. Again, the conflict is not between Christianity and science, but between Christianity and the philosophical system of naturalism, or materialism.

The Church's Role in the Conflict
Just as scientists have been guilty of making unjustified theological pronouncements, so also theologians have at times ventured ill-advisedly into the arena of science. For example, the question of whether the sun revolves around the earth is one for science, not theology. Unfortunately, this realization came too late for Galileo, whose only crime, to paraphrase one contemporary songwriter, was searching for the truth.

The response of the church to the cultural success of naturalism (or humanism) has been a bit of a paradox. Prior to the arrival of naturalism, theology had considered science at least a junior partner. Theologians often wanted the last word, but they valued the support given to theological ideas by scientific concepts, for ultimately they believed that all truth was one.

By the nineteenth century, theology felt that it had gotten beyond the mistakes of the past. It knew the painful consequences of clinging to an outmoded scientific theory that had wrongly become wedded to theological truth—such as the continued support for Ptolemaic "earth-centered" astronomy when it was challenged by Galileo's support of Copernicus's theory that the earth orbited the sun. For instance, as we will see in the next chapter, most theologians did not view the new science of geology as a serious challenge to Christian truth.[6]

But by the twentieth century, the world was turning dark and frightening. The Scriptures were under attack in the churches. Students were losing their faith in the schools. Bolshevism was on the rise in Russia. Germany had attacked Europe. And in industry capitalism was grinding down the working poor. In all of these movements popularizers claimed the support of science, under the name of "evolution." What was the "great commoner," William Jennings Bryan, to do? His crusades against the "evolutionary foe" led straight to the Scopes trial in 1925, in Dayton, Tennessee.[7]

The greatest effect of the Scopes trial, however, came not from the original judicial carnival, but from the mythic retelling of the story in the play *Inherit the Wind*. The hit Broadway play recast the Scopes trial as a precursor case of McCarthy-era repression of free speech by the reactionary, ignorant fundamentalists.[8] Small wonder that the church felt that it was being deserted by mainstream science in favor of the naturalism philosophy.

But rather than rejecting science entirely, the church looked for new scientific partners. The problem with this partnership was that many of the new scientific experts adopted by the church were not experts at all. (For instance, "young earth flood geology" can be traced to the "revelations" of Mary Ellen White, founder of Seventh-Day Adventism.[9]) Resolutions put together by such "experts" have tended to fall apart in the face of the mountains of data collected from nature by scientists, including Christian scientists. In reaction, rather than accepting science, some in the church have chosen to ignore it, thus giving rise to admonitions such as "Do not study science; you will lose your faith."

Individuals who point out the problems of any of several popular "faith-science" integration packages are accused of abandoning their faith. For example, in some Christian circles those who state that a "young earth" is not supported by the physical data are accused, Galileo-like, of denying the authority of Scripture. But to insist on hanging on to such outdated science in the face of overwhelmingly opposing physical evidence is to give theology authority not over science, but over nature. And only God has that authority.

God's authority is supreme. As we noted earlier, both Scripture and nature derive their authority from God. Both are God's revelations of truth. The Scriptures describe nature as God's obedient creature, speaking truth (Psalm 19:1-3). Thus, any theologian who ignores or denies the facts of nature is, in essence, placing human authority above God's authority in *both* areas. The idea that theology has authority over nature actually denies the authority of Scripture, which, again, holds a high view of nature as testifying to truth.

Most of the time, the conflict between science and theology is a result of the "literalism" error that we noted earlier. Based on a superficial glance, the reader (of Scripture and/or of nature) forms a strong opinion of the meaning. The motive to defend God's truth (or to destroy error) is commendable, but a good motive is not enough to save a wrong idea or a conviction that flies in the face of the physical evidence. Often, the proponents of wrongheaded notions know far more than their listeners, but far less than they think they know. Some may not be aware of their errors, while others may intentionally omit certain facts in order to make their case seem stronger. But an honest teacher has the obligation to search out all truth and not conceal it. In a very real sense, my purpose is just that: to search out and to explain the truth in the controversial area of evolutionary thought. Believing that all truth is God's truth, I also make it my goal to search with faith and to seek the resolution of these controversies.

CHAPTER 4

The Age of the Earth

S CIENCE BEGINS WITH THE OBSERVATION OF PATTERNS. EVALUATING the credence of various ideas and tenets associated with evolution will entail observations related to things such as physical processes, rocks and fossils, and living creatures.

Focusing on the age of the earth will provide a good and fairly simple and straightforward example of how the scientific process works—of how scientific methods lead to conclusions or tentative conclusions about specific aspects of nature. For some, the age of the earth relates directly to theories associated with evolution. For example, if according to evolutionary theory a particular change requires at least a million years, that theory cannot be true if the earth is only about ten thousand years old.

"Proving" Past Events

How do we prove history? How can we know about events that we were not present to witness? One commonsense approach is to find a process that is still going on today and look for evidence that the same process occurred in the past. For instance, if we cut down a tree, we would notice rings of light and dark wood in the trunk. If we had never seen a cut tree trunk, we might not think of the rings as evidence related to time. To us, they would merely be stripes like those of a tiger.

However, a little experience in the woods—watching trees getting thicker and taller—would quickly lead us to the realization that tree rings are actually growth rings. Further observation would lead us to conclude that they are yearly rings. We would come to know that there is a wide light ring resulting from rapid growth during the balmy summer, and a narrow dark ring that forms during the cold winter. We then would know that we could count the rings to determine the age of the tree.

If we matched the width of the rings with our memories of what the weather was like over the last few summers, we would be able to conclude that wider light rings formed during good summers, while narrower ones formed during poor summers (perhaps a drought). Assuming that this was a very old tree, we would then feel some confidence in concluding that the summer of 1753 was a very dry year, because the ring for that year was narrow. We cannot go back and see the summer of 1753, but we can predict (or "postdict") that year's weather based on our "theory" of a dry year. We could predict that any other tree in the area that was alive in 1753 would also sport a narrow ring caused by the drought of 1753.

Based on the ability to determine the age of the tree, we could also rule out various other histories. For instance, if someone theorized that between 1820 and 1860 the entire area around the tree was a lake twenty feet deep, we would, with confidence, disagree. This idea could be proved wrong, for the tree must have stood at that spot throughout those decades, and it could not have lived in a lake. None of this is complex; it is merely common sense.

Science has come to understand many natural processes that have regular rates, similar to tree growth, rates that can be evaluated to date objects in nature. The world has many "clocks." Here are a few: Yearly layers are laid down by flooding rivers and by the snow falling on glaciers. Radioactive materials decay at precise rates. Flints and tooth enamel collect photons in their crystals. Neutral mutations accumulate in DNA. Molten rock cools at specific rates. Often more than one "clock" can be applied to the same phenomena, allowing them to cross-check one another.[1]

The existence of such natural clocks brings up issues related to the Scriptures. First, does the Bible, properly read and understood, really require a date for creation only a few thousand years ago, for a universe that starts only a few days before the first human beings? Second, does nature require a much older date for the creation, with humans arriving relatively late? What does science have to say about how old things are?

What Science Tell Us

We will look first at an example of the evidence from nature. Then we will consider how the evidence might be understood in light of the Scriptures.

The "tree" that I want to look at is a coral reef. Coral consists of the leftover exterior "skeletons" of polyps, a group of animals related to jellyfish and sea anemones. Coral polyps extract minerals from the tropical seas. Then they secrete those minerals in the form of a rock-hard exterior "cavity" within which they live. They must live at a depth between one hundred and three hundred feet to have enough light to grow. They live in very large colonies. Each generation grows on top of the skeletons from the previous generations. As the years pass, coral colonies can grow into enormous masses. However, they do not grow rapidly. The fastest coral growth ever measured was two-fifths of an inch per year, which is about one foot every thirty years. Thus, the evidence from nature suggests that even if conditions are perfect, a reef can grow no more than thirty-three feet in one thousand years.

The first hydrogen bomb was tested at Eniwetok, a coralline atoll. An atoll is a ring-shaped island formed by coral growing on the slopes of an underwater mountain. Before the bomb was detonated, a hole was drilled through the coralline cap of the island, down to bedrock. The drill went through 4,610 feet of coral before it hit the volcanic rock of the mountain. That much coral would take about 140,000 years to grow, even if growing conditions were perfect. Based on the historical events recorded in the core sample, we know that in several ways conditions were not perfect, so the Eniwetok reef is actually much older than 140,000 years.[2]

Obviously, a 140,000-year-old coral reef cannot logically exist on an earth that is only about eight thousand years old. But we know that the reef exists. It is exactly like the 300-year-old tree that proves that no lake existed at that site in 1830. And there are many other examples of such "old" objects in nature. For instance, the magma (molten rock) that formed the palisades of New Jersey would have taken one hundred thousand years to cool.

To go into all the details of how scientists know what they know is beyond the purview of this book. Suffice it to say that the scientific community (believers and nonbelievers alike) is overwhelmingly confident that nature's clocks indicate an old earth with a long history, a history that offers evidence that the earth has changed over time.[3]

The Earth's Pattern of Change

What did the ancient world look like? We first must acknowledge that "seeing" the past is to some extent a chancy thing, as we are limited in what we are able to see by what has been preserved. Scavengers almost immediately destroy most things that die. However, sometimes unusual conditions provide a sort of "snapshot" of a particular era. For instance, there are three such clear windows on a particular ancient time, 544 to 490 million years ago, called the Cambrian era (part of the Palezoic), an era when, according to the evidence, animal life was just beginning. These windows are the Burgess shale in western Canada (518 mya [million years ago]), the Chengjiang shale of China (532 mya), and the Sirius Passet shale of Greenland (534 mya). In these fine-grained mudstones we can see not just tough shells, but hundreds of beautiful fossils showing the entire bodies of fragile sea creatures. These are records of unparalleled completeness, clarity, and beauty.[4]

The Cambrian seas apparently were quite different from those of today. They show no creatures much like anything now alive. Many were very odd indeed. For instance, most echinoderms (starfish, for example) had three-part rather than modern five-part bodies. Many of the arthropods ("jointed foot" types like crayfish and roaches) do not fit into any modern groups. Anomalocaries and Opabinia, with their odd appendages and tail fins, hardly seem to be arthropods at all.

There were no fish with jaws, as there are today. However, there were creatures that swam, floated, crawled, plowed the surface, and burrowed under it. There were large, heavy creatures and tiny, fragile ones. We can recognize predators and their prey. Almost all the major animal groups called phyla were represented, but none of them by their modern forms. And, again, some of the Cambrian creatures cannot be grouped with any forms living today.

What do these fossils show? These were creatures largely unlike creatures that we see today, but obviously they bear a relationship to modern creatures, remote though it may be. They lived in a complex, integrated biological world, playing the same roles as do the animals of today. They were worldwide in their distribution; today's animals simply were not yet in existence.

What, then, does nature report of the history of the earth? Let us take a quick trip to the front end of the universe. Everything of which we are aware—time and space, matter and energy—began about thirteen billion years ago as an unimaginable explosion of light. This conclusion is based on a great deal of complex astrophysics, including background radiation and the redshift of the light from distant galaxies, explanations of which are far beyond the scope of this book.

This explosion event, commonly known as the Big Bang, is not explained by science. Many researchers certainly are trying to find a plausible cause for it, but at this point, the Big Bang is best understood as a description of creation *ex nihilo* (from nothing). That is, it has no scientifically proven cause, and perhaps no provable cause.

The current model of science divides time (the thirteen billion years) into six major periods or eras.[5] The first, which is about two-thirds of the total, may be termed the Cosmic era. During this period stars were born, grew old, and blew up. In this process the stars forged the elements (such as carbon) necessary for life (as we know it) in their internal furnaces and released them. This enrichment made the next generation of stars and planets capable of supporting life.

The Archeozoic era began five billion years ago as the earth formed, followed by the formation of its seas, atmosphere, and continents. This era lasted about two-thirds of the remaining time. Almost as soon as the earth got cool enough (around 3.8 billion years ago, as determined by radioactive-decay "clocks"), the first simple living things (bacteria) appeared in the sea. The fossil evidence includes both visual and chemical traces of the bacteria.

The Proterozoic era began around 1.7 billion years ago and lasted two-thirds of the remaining time. During this period the complex (eukaryotic) cells appeared, cells from which plants and animals are made. The green algae formed during this period began releasing oxygen into the air, though nothing yet lived on the surface of the land.

The Paleozoic era began with what is known as the "Cambrian explosion" of animal forms around 530 million years ago. Again, it lasted about two-thirds of the remaining time. The Paleozoic era also saw the earth's ecosystem change from one dominated by invertebrates (such as trilobites) to one dominated by vertebrates (fish and amphibians). Animals and plants became established on dry land. However, this era ended with a major disaster, one that destroyed 96 percent of the species on earth.

The Mesozoic era probably makes you think of dinosaurs. The Jurassic period, popularized by filmmaker Steven Spielberg, is part of the Mesozoic, which began about two hundred million years ago and lasted two-thirds of the remaining time. During the Mesozoic, the first mammals, birds, flowering plants, modern fish, and amphibians appeared. Indications are that this era ended when a mammoth asteroid struck what is now the Yucatan Peninsula of Mexico, destroying most living creatures, including all the dinosaurs. Why is this the prevailing conclusion? Because there is a very narrow iridium layer in the rock strata worldwide. Such layers are caused by vaporized rock from a large meteorite strike. Below this layer, dinosaur remains are plentiful; above it, there are none.

The sixth era, the Cenozoic, in which we live, began sixty-five million years ago. It is characterized by the dominance of mammals and flowering plants. Human beings appeared toward the end of the Cenozoic, but this is a topic for a later chapter.

Does this description of creatures appearing over vast amounts of time somehow mean that God did not create them? Or does it mean that God was unable to create them all at once? Quite the opposite! God does not need time to create. In fact, it is God who, acting from eternity, creates time itself. Thus, when God creates creatures, he creates their whole history (Psalm 139:13-16). God does not need a long time to create. But God can, if he chooses, create a long history. The data suggest that God did exactly that.

The earth has changed dramatically since the days of the first animals. How and when was this strange past discovered? By the middle of the seventeenth century, the science of geology began to uncover evidence of "deep time" and strange creatures. The present geological timetable was put together by the first half of the nineteenth century, based on a predictable series of distinct fossil groups. The deepest rocks held the fossils that were the most different from the present.

This investigation was not carried out to prove Darwin's theory. In fact, Darwin had not yet published, nor even started to think about species. One motivation for digging, unglamorous though it may be, was industrial: they were looking for coal. The Carboniferous layer contained enormous quantities of plant fossils—coal. The fossil trees were different from today's plants, more like modern horsetail rushes and ferns.

As the railroads and canals were cut through the hills, surveyors such as William Smith noted that the Carboniferous layer was above certain other layers (such as the Silurian) and below others (such as the Devonian). Each layer had its own unique group of fossil organisms. The reader who knows British geography will recognize the names of these layers. Each layer was named for the location where the layer is found on the surface. So if you found fossils of the Devonian assemblage on the surface, you could dig down and find the coal layer. If Silurian fossils were on the surface, you were already below the Carboniferous, and you could "dig to China" without finding coal.

At first, the church encouraged these geologists, most of whom were Christians. In fact, many were ministers. For instance, Rev. William Buckland, geologist at Oxford University, was commissioned by the Earl of Bridgewater to write an extensive treatise using science to defend the faith. His books, published in the 1820s, used the geological record of biological change to defend the Scriptures. Throughout the nineteenth century, conservative, Bible-believing Christians held most of the major university chairs of geology in Britain and the United States.[6]

Obviously, these Christian geologists did not think, as do some today, that the first chapter of Genesis meant a single week of time. But what did the theologians think? As we noted previously, most of the major theologians of the time accepted an old earth—that is, an ancient creation. They interpreted Genesis in one of two ways. According to the Age-Day interpretation, each "day" was a long, creative period. According to the second interpretation, the Ruin-Reconstitution or Gap theory, the geological ages described above are inserted between the first two verses of Genesis.

These theologians included Delitsch, Lange, MacLaren, Hodge, Shedd, Orr, Strong, Warfield, Miley, and others. Though these are not exactly household names in the church today, these men were leading Presbyterian, Baptist, and Methodist theologians of their time. Some wrote commentaries that are still used for sermon preparation today.

These two positions continued to be the most common conservative interpretations of Genesis through most of the twentieth century as well. This was especially true within American fundamentalism, under the guidance of the notes in C. I. Scofield's reference Bible.[7]

What is the "moral" of this story? Simply that an old earth posed no theological problem for the church of an earlier era. However, for some theologians today, the problem is that a very old earth is not what the Scriptures seem to be describing. The discrepancy first was faced during the eighteenth century as geology began to uncover evidence of great age. Around 1650 certain scholars (notably, Bishops Lightfoot and Usher) had applied a "scientific" analysis to the genealogies in the Scriptures, compared them with secular sources, and concluded that the earth began in 4004 B.C. As evidence accumulated that the earth was far older, there was some concern in the church. The major concern was the use made of the new geology by anticlerical figures of the Enlightenment to attack the authority of the Scriptures.

The typical response of conservative theologians during the nineteenth century (men such as Augustus Strong, a Baptist, and Charles Hodge, a Presbyterian) was that Lightfoot, Usher, and others had incorrectly interpreted Genesis, had failed to understand that the "days" of creation were long periods of creative activity (the Age-Day interpretation). Most American fundamentalists, however, followed the ideas of C. I. Scofield (originally proposed by Thomas Chalmers of the Free Church of Scotland in 1790). Scofield's notes place the geological ages in a "gap" between the first two verses of Genesis.[8] (As we noted, these two old-earth interpretations of Genesis dominated the conservative church in the United States until the early 1960s.)

One response to old-earth evidence worth noting, since it continues to crop up, was the proposal of Philip Grosse that God created the earth with the *appearance* of age. According to this theory, the earth was only some four thousand years old, but it appeared to be much older. If God did that, it was indeed possible to have an "old" coral reef on a young earth. But we also could not prove that a tree was 250 years old. Perhaps God made the tree in such as way that when it was one hundred years old, it had the appearance of being three hundred years old. In fact, you could not even prove that I borrowed your money last year. Maybe God made the world last month, complete with your memories of borrowed money!

It seems to me that we are stuck. If God writes fiction, then we cannot know anything about reality at all. I prefer to posit that God tells real history rather than fiction in nature: "The heavens declare the glory of God; and the firmament sheweth his handywork. Day unto day uttereth speech, and night unto night sheweth knowledge" (Psalm 19:1-2, KJV).

After 1960, however, the most common interpretation of Genesis in the American church changed to young-earth/flood geology. This interpretation had also appeared during the nineteenth century, notably in the extrabiblical "revelations" of Mary Ellen White, founder of Seventh-Day Adventism. White's writings, considered binding on the SDA church, advocated a recent creation and a universal flood, as well as Saturday worship and dietary restrictions. Her view of creation was popularized during the 1930s in a series of little books written by

George McCready Price (a Seventh-Day Adventist and self-taught "geologist").[9]

Price's disciple Henry Morris crusaded for this young-earth interpretation in his 1961 book (with John Whitcomb) *The Genesis Flood*.[10] Crusaded? Yes. Using the basic rhetorical device that to disagree with this interpretation of Genesis was to deny the authority of Scripture, the Creation Research Institute and other organizations effectively declared the views of Scofield, Hodge, and Strong heretical. A strong word? Indeed it is, for no person has the right to declare his or her own *interpretation* of Scripture to be infallible.

To conclude, the "problem" of the age of the earth is ultimately a conflict not between science and Scripture, but among theologians. For as we have seen, theologians who are equally committed to the authority of Scripture have disagreed on the question of the age of the earth. The young-earth view is not necessarily the majority view among conservative theologians, but it has become a "hot button" topic that many tend to avoid. Unfortunately, that avoidance allows misunderstandings to be taught without challenge. Suffice it to say that accepting evidence for an old earth should not necessarily be associated with evolution, whether godless or not. This was not the case in the past, and it should not be the case today.

CHAPTER 5

Defining "Evolution" and "Creation"

IN LATER CHAPTERS WE WILL EXPLORE WHAT WE KNOW (BASED ON observing God's nature) about the origins and development of life, including human life. Unfortunately, the process of simply exploring nature and reaching whatever conclusions it leads us to has to some extent been corrupted by those who have made claims outside the boundaries of science or theology, claims that have had a polarizing effect. The very reference to the "creation-evolution debate" reflects this polarization.

I have trouble understanding how anyone can respond with a simple yes or no to the common question "Do you believe in evolution?" It should be obvious that the answer to that question depends on how evolution is defined or understood. One person's definition may be very different from another's. In fact, both "evolution" and "creation" are words that have been overloaded with multiple meanings. Does the word "evolution" refer to any change at all to living species? Does it have to do with the physical mechanisms (causes) and processes that lie behind the change? Does it have to do with progress? Biological relationships? To some, the idea of evolution and the idea of a Creator God are mutually exclusive concepts. To others, these ideas are perfectly compatible.

We cannot simply ask whether evolution is true or false, or whether it is scientific or not, or whether it is anti-God—for the answer to these questions invariably will depend on what the word "evolution" means to the person who is responding to these and other questions. If "evolution" means that one does not believe in God, most people would say no. But if it means that wolves and dogs are biological "first cousins," most would say yes. So before determining how we might evaluate this thing called evolution, we must first unpack its specific meaning or meanings.

Another reason to begin with definitions is so that we can fairly and clearly frame the discussion—agree on the rules for the debate, if you will—between and among those with varying viewpoints. By changing or smudging word meanings, arguments can be dodged or forced in certain directions. Thus it is best to be clear from the start by defining terms.

Evolution as Change

Sometimes all that a person means by "evolution" is change (as opposed to stasis) in a biological lineage. As generations pass, groups of animals and plants may show fast change, slow change, or no change (stasis). The change could be steady, oscillatory (back and forth), or intermittent. It might be directly observed, or it might be deduced from fossil evidence. Specific analysis of change is based on careful observation.

Keep in mind that the extent and type of change does not speak to how or why the change or stasis might take place. It says nothing about the mechanism (the cause) or about possible goals or purposes for the change.

Should change in groups of organisms be considered evolution? Or might the same kinds of changes be associated with creation? As we consider evolution as change, the irony is that ordinarily one would consider change to be an aspect of creation. After all, creative people are people who change things by making them different, new. In early Christian thought, stasis, or the idea of an eternal (uncreated) universe, was considered to be a secular Greek concept. Change, on the other hand, was to be expected in a world once created and eventually coming to an end. But today, of course, things have been turned around. Evolution has become a synonym for change, and (biological) stasis is associated with creation. In and of themselves, however, neither stasis nor change says anything about cause or intent. They are simply descriptions of observed reality.

Evolution as History

History is essentially the reconstruction of events and realities of the past. It is in part the objective reporting of what the data show. History is also partly reconstruction and analysis. With regard to our topic, historical reconstructions typically are termed either "evolutionary" or "creationist," depending in part on how they fit within some larger philosophical or theological framework.

However, again because "evolution versus creation" is too simplistic a dichotomy, specific reconstructions of history do not fit neatly into either a creationist or evolutionist category. For instance, although a young-earth view is held by some creationists, many who hold an old-earth view have, since early in the nineteenth century, also called themselves creationists. Thus it would be more accurate and fair to consider "young earth" and "old earth" as possible histories, but not to associate them exclusively with either creation or evolution.

The hastiness with which some people label others as creationists or evolutionists has clouded the debate and hampered a reasoned pursuit of truth. Take, for example, those who agree with some aspects of evolutionary theory but ultimately consider themselves creationists. To tag someone who thinks that way as an evolutionist without qualification or definition often is meant to suggest that the person is wicked or foolish, or perhaps both. This is an easy but dishonorable way to discredit someone who disagrees with you, and it is especially unfortunate when the "opponent" agrees with you 90 percent of the time. The truth is that many who ultimately consider themselves to be creationists disagree among themselves on some specific theological and scientific interpretations. The same is true of so-named evolutionists.

Evolution as Radical Change

Descent with modification is one of the most familiar concepts associated with evolution. It means in essence that present-day organisms are descended from significantly different ancestors. But how much difference constitutes "significant"? Obviously, none of us are clones of our parents, so we must be different from them. But few people would say, "I have 'evolved' from my father." Thus, discussing the nature of evolution as relationship (or descent with modification) requires rules for deciding which "types" are truly different from others.

It also requires a definition of "descent." Presumably, descent means offspring produced by a regular pattern of reproduction and development. But are mutations (sudden changes in the genes being inherited) to be included as a part of "normal" descent? What about cloning and/or the adding or changing of genes by a genetic engineer? We cannot assume that such events are always due to human intervention alone. They may in fact occur without human direction—for instance, by what is known as the "lateral transfer" of genes by an infecting retrovirus.[1] Most would still consider such a genetically modified offspring to be descended from its parents.

But what if such genetic engineering is due to "the hand of God" adding a gene (either directly or by using some "natural" mechanism)? Would divine genetic engineering be called descent with modification or not? Would it be a form of creation? Might it be both? If you are convinced that the deity does not do evolution, then you will say that it cannot be both. This is the opinion of both Richard Dawkins (evolutionist) and Henry Morris (creationist). Not everyone agrees. There are those who say that it can indeed be both. A related distinction is often drawn between macroevolution and microevolution. Both terms refer to descent with modification. Microevolution refers to descent with modification at a very limited level—for example, the change seen over a dozen or so generations descended from a particular individual. Macroevolution, a more far-reaching concept of descent, is the idea that all organisms are descended from a single, ancient ancestral population.

There is no clean line by which to tell where "micro" ends and "macro" begins. Everyone accepts some level of descent with modification. After all, the traditional Christian view is that all humans are descended, with modification, from a single ancestral pair. But how extensive can biological change be and still be considered "micro"? There is no easy answer, because ultimately the distinction is arbitrary, open to debate.

In sum, the expression "descent with modification" refers to groups of organisms related to one another by descent from common ancestors, groups that have become different as the generations have passed. Everyone (including creationists) accepts this concept, within limits. The question is whether there are limits. And if there are limits to what can unarguably be considered descent with modification, what are they and by what principles are they established?

Evolution as Certain Forces of Nature

Certain forces of nature act to change, or to prevent change, in populations of animals. These forces can explain much of the observed changes in populations, and thus they are considered mechanisms (causes) of evolution (or of stasis if they are preventing change). Such natural forces have come to be associated with the term "evolution."

The best known of these natural causes, or mechanisms, is known as "natural selection." The concept of natural selection is rooted in the observation that not all organisms are equally adapted to their environments. Some have higher fertility or viability, and thus they leave more offspring than the less fertile or viable ones. Over time this difference changes the average character of populations. In agriculture this process is termed "breeding" or "artificial" selection. In nature it is known, again, as natural selection.

Another natural force that changes populations is the phenomenon of genetic drift, sometimes termed "neutral evolution." Genetic drift refers to a change in the average character of a population due simply to the specific characteristics of the organisms that happen to be born each generation. The characteristics of a population may also be shifted by the immigration of individuals with genes different from those of the resident population. New genes (usually variants of old ones) may also be added through mutation or recombination (crossover between chromosomes). And sociological developments can arise that isolate parts of a particular population, giving it the opportunity to become different in some way from the broader population. For example, Amish people have a higher incidence of fused fingers because they (and their genes) are to some extent isolated from the broader population.

Note that all such processes are simply nature at work. They are typically and collectively referred to as "evolutionary mechanisms." The existence of these processes cannot be doubted, but their efficacy is another matter. To put it another way, the question is, "Are the forces of nature able to produce change or prevent change on their own?" The answer to this question depends on one's view of nature. Is nature an autonomous machine, or is it governed by God? This is not a question for science to decide.

Evolution and Design
Typically, design is considered to be the essence of creation. However, certain *positions* on design are frequently associated with evolution. One example is the view that the structures and

products of nature have no design and do not follow any rational blueprint. According to this position, adaptations may be contingent on happenstance or on the laws of nature. Such views clearly are philosophical (or religious) positions about "teleology," defined as the philosophical study of design or purpose in natural processes. Advocates of this position frequently argue that their convictions are necessary conclusions based on the data supplied by nature. They insist that the efficacy of natural mechanisms makes them necessary truths.

But this begs the question. That is, this conclusion is based totally on the beginning philosophical presumption, the presupposition. Since those who hold this view have committed themselves to view nature's mechanisms as completely autonomous, and by doing so have assumed God's absence, it is no surprise that they can find no ultimate purpose in nature. They then call their philosophical position on teleology (purpose) "evolution." In this instance, "evolution" has been defined essentially as a "blind watchmaker."[2]

Evolution as Cosmology

This category of evolutionary "definitions" is related to the previous one. It refers to worldviews, or *Weltanschauungen*. Several different views of reality as a whole have been associated with evolution at one time or another. One of the more familiar ones today is simple materialism, the belief that only matter exists and that everything we see is merely a result of the properties of matter. According to this cosmology, the world is a product of chance and necessity; it has no direction or purpose. As noted earlier, this view dates back several thousand years to Democritus. Contemporary materialists such as Richard Dawkins, the heirs of the naturalism movement, as well as many creationists, tend to identify this worldview with the term "evolution."

But materialism is not the only "evolution-as-cosmology" variant. The more humanity-centered views of the Enlightenment era placed human progress as the central fact of reality, and they also frequently identified this assumed progress principle with the word "evolution." In fact, this central myth of Western society—the myth of inevitable progress—is so identified with the word "evolution" in the popular mind that many use "evolution" simply as a synonym for "change for the better."

According to both the materialistic and the humanity-centered cosmological positions, the universe and humanity are autonomous, undirected by any influence from outside the natural world. Keep in mind, however, that during the nineteenth century most people viewed evolution as a directed process. It might be "intrinsic" evolution due to innate internal forces (sometimes termed "vitalism"), or it might be "extrinsic" evolution due to exterior guidance (presumably extramaterial forces or entities).[3] Keep in mind that the Enlightenment viewed God, if at all, as an impersonal being or force. More recently there has even been a sort of "divine" evolution proposed, according to which God himself is evolving and bringing the universe along. This position is often identified with the ideas of Teilhard de Chardin.[4]

Again, the concept of change is by no means foreign to Christianity. Since the beginning of the Christian era, some have held that external guiding forces direct natural development, and that these forces are to be identified with the personal God of the Bible. If the forces (God) are deemed to have acted only at the beginning (preprogramming), then the view is termed "deistic"; if the forces (God) are held to continue to act throughout the process, then the view is termed "theistic." Thus we may also have various theistic or deistic "evolutions" as forms of evolution as cosmology.

Evolution as Paradigm
In 1973 Theodosius Dobzhansky, one of the major architects of the new Neo-Darwinian Synthesis, authored a paper entitled "Nothing in Biology Makes Sense Except in the Light of Evolution."[5] By this,

Dobzhansky meant that evolution was a "paradigm," a key idea that made sense of, and gave unity to, the diversity of the world of life. Thus biology would be, for all intents and purposes, the science of evolution. Therefore (the logic goes), to reject evolution is to reject all of science, all modern medicine, the "Green Revolution" in agriculture, or even the use of computers.

Of course, this is nonsensical, or at least very simplistic. The problem with calling evolution the focal point for biology is that the meaning of the statement depends on what the speaker means by "evolution." And since, as we have seen, different people mean different things by the word, the meaning of the "evolutionary paradigm" changes, based on who is defining it. It might mean that the structures of modern organisms are best explained by presupposing a historical process of descent that caused them all to develop gradually from a single ancestor. Therefore history would be the key to the present. Another person might mean that the processes and adaptations that we see are best explained as adaptations to the environment, adaptations produced as natural selection acts to change the genomes of populations that are interacting with their environments. Therefore adaptation becomes the key to the present. Or evolution as paradigm could mean that the proponent believes that all living things have been produced without any design or purpose, as multiple accidents, and thus all biological knowledge must exclude statements of purpose. Thus biology becomes dependent on the investigator holding the "right" philosophy. (As a member of the Russian Orthodox Church, it seems unlikely that Dobzhansky would mean that!)

Confusin' Evolution

By now it should be clear why the word "evolution" causes so much confusion. It has been used by different people to mean different things at different times. Patterns of change, events of history, natural mechanisms, genetic relationships, disciplinary paradigms, questions of purpose, and even positions that address the very basis of reality have all been associated with evolution. Obviously, these various associations need to be considered independently.

Nor is the use of the word "creation" much clearer. It too has been defined by terms that fit into these various categories. For instance, is creation a theological principle that God created all things? Is it a historical scenario that the earth is only a few thousand years old? Is creation a theory that natural, biological mechanisms are incapable of producing significant change? Is it a proposition that plants and animals look essentially the same as those originally made? Is creation a paradigm for biological classification stating that the fundamental biological unit is the "baramin" (a term invented by contemporary creationists from the Hebrew words *bara*, "create," and *min*, "kind")?

In sum, clear and honest thinking requires clear definitions. Blurry definitions lead to confusion, and sometimes to deception. It is rather easy to work a logical "shell game" (with truth as the hidden pea) when so many things are confounded. Beware of the person who attempts to tarnish either the word "evolution" or the word "creation" by using the tactic of guilt by association. Consider the person who associates "evolution" with atheism and then argues that opposing atheism requires believing that the earth is only a few thousand years old. Those on the other side do the same thing when they associate the word "creation" with religious fanaticism or simplemindedness and then argue that rational science requires dropping the "God hypothesis."

Such tactics have nothing to do with logic or truth. They are unfortunate, given that the opponents in this debate may have more to learn from one another than they realize. For as we have seen, depending on how they are defined, the terms "evolution" and "creation" are not as mutually exclusive as some have assumed.

CHAPTER 6

Understanding Cause and Effect

A MIDST THE VARIOUS WAYS OF UNDERSTANDING EVOLUTION and creation, the concept of "cause and effect" provides a useful framework for sorting them out. Aristotle distinguished between and among four types of cause, all of which apply to anything produced. According to his framework, *final cause* is the intended purpose of the maker of an object. *Formal cause* is the maker's plan or blueprint for the object. *Material cause* is the raw material from which the object is made. *Efficient cause* is the force applied to the raw material that actually produces the object. A full accounting of the existence of any object thus requires a description in four dimensions.[1]

For example, consider the shape of a flowerpot. The potter plans to make a pot suitable for growing flowers. This is the final cause, or the purpose of the flowerpot. The potter chooses an appropriate dense clay—the material cause—the raw material that will be shaped into the flowerpot. The potter chooses a pot design—the formal cause or blueprint—to guide the process. And then, with clay, design, and purpose, the potter throws the pot onto the wheel, shaping it by hand and with tools—the efficient cause, or force(s) applied to the raw material. Again, the complete "cause" of the pot's makeup is fourfold.

When applied forces produce human artifacts, such as lawn

rakes and orbital shuttlecraft, there is no denial of purpose. That is because as we humans make tools, the shaping forces (efficient causes) acting upon the raw materials (material causes) are under the direction of our intelligent design or blueprint (formal cause) chosen for the tool that we intend to make (final cause).

But is this also the case with nature? Does the concept that efficient causes (applied forces) have brought about such varied things as aardvarks and roses, humans and sequoias, somehow deny the possibility of purpose? Clearly, it does not. Aristotle's four causes remain meaningful. For objects produced in and by "nature," it is still the case that efficient causes (forces) act on material causes (raw material) under the "direction" of formal causes (blueprints). But what are nature's blueprints? In both human and natural causality formal causes act as "boundary conditions," the specific physical states that constrain or direct the forces of efficient causes. To put it another way, the forces that act upon raw material in nature can be governed, or directed, by things that can act as blueprints. An example of such a physical formal cause, or blueprint, is the surface of a growing crystal. As the crystal grows, new material attaches in a regular pattern determined by the existing surface. Think of how ice crystallizes on a cold window to form "vines."

Perhaps the best example of a formal cause (blueprint) in nature is the pattern of bases on a strand of DNA. Such blueprints in nature set limits, establish boundaries on what can happen when forces act on raw materials. Once identified, these boundaries free those who observe nature to make predictions and draw conclusions. Without such boundaries, all developments would be random, unpredictable. And as we noted previously, under those conditions the very idea of doing science would be absurd.

But what about final cause, or purpose, in nature? Far from being off limits, the question of final cause cannot be avoided. For those of us who believe that there is an ultimate purpose for creation, such a conviction confirms the reality and importance—that is, the relevance—of the material and efficient causes studied by science. Intelligence (i.e., divine intelligence) acts as the source of formal causes in nature, determining the natural boundaries that shape nature's direction, and thus providing purpose. To quote a convinced "teleologist," Princeton theologian B. B. Warfield, at the beginning of the twentieth century, "Some lack of general philosophical acumen must be suspected when it is not fully understood that teleology is in no way inconsistent with—is rather necessarily involved with—a complete system of natural causation. Every teleological system implies a complete 'causo-mechanistic' explanation as its instrument."[2]

But of course, an intelligent designer is not the only possible answer to the question of final cause in nature. Consider the view of G. G. Simpson, a staunch "antiteleologist" (someone who believes that there is no purpose) born half a century later than Warfield: "It is already evident that all the objective phenomena of the history of life can be explained by purely naturalistic, or in the proper meaning of a much abused word, materialistic factors.... Man is the result of a purposeless and natural process that did not have him in mind."[3]

Note that this was not the view of a man who considered *questions* of purpose meaningless. Rather, Simpson viewed the material cosmos itself as the adequate and autonomous final cause. According to this view, humanity ultimately can have no purpose to its maker, for its maker (autonomous matter) has no intentionality or sense of purpose. Significantly, however, Simpson does address the *question* of purpose. Even a materialist cannot deny that there is meaning to the *question* of purpose (regardless of how it is answered), for such a denial would imply that the materialist's own statements denying purpose are themselves meaningless nonsense.

Ultimately, the most significant differences between the various evolutionist and creationist positions have to do with differences over the question of final cause, or purpose. Questions of material or efficient cause (raw materials and the forces that act on them) related to how species have changed or whether there is evidence of a species of some prehuman "caveman" are questions that can be answered, one way or another, by observation, by studying and analyzing the evidence, including the genetic evidence. But the ultimate question is this: what or who is directing the changes?

Simply put, it all goes back to distinguishing between the realm of science, which focuses on observation of the natural order, and that of theology, whose domain consists of beliefs and convictions related to purpose and to who and what lies behind the natural order.

You might "believe" that the basketball hoop is ten feet high, and I might "believe" that it is twelve feet high. But this is not a question of belief. It is a matter for observation. We can measure the height of the hoop and determine who is right and who is wrong. Similarly, questions pertaining to species formation and the origins of human beings are not ultimately questions of belief; they are subjects of scientific observation.

To be sure, there is a genuine struggle, a real motivation to win the allegiance of others via the argument. But the fight is not exactly about evolution, as it is commonly understood. People might very well be in agreement on some of the ideas associated with evolution, while ultimately residing in diametrically opposed philosophical camps.

Intelligent Design
The latest arguments developed to oppose evolution focus on the question of intelligent design. The idea of intelligent design is, of course, directly related to the question of whether the universe has an ultimate purpose. Every thoughtful person holds some position on this issue. People think either that some intelligent being or force designed the universe or that no intelligence shaped it, or they remain agnostic, unsure as to the correct answer. A person's

convictions on teleology (purpose) contribute greatly to his or her perception of design in nature.

No one who believes in a personal God denies design and purpose in nature. But we must acknowledge that God can choose to act in any of a number of different ways to realize those purposes. Thus, belief in design can be coupled with a wide variety of ideas about secondary (efficient and material) causes and historical events. "God believers" can disagree considerably on God's methods, and therefore they can disagree on what sorts of evidence would actually display God's hand.

On the other hand, people who believe in autonomous matter must accept matter itself as the ultimate shaping force, the equivalent of a "designer." Material cause becomes the bedrock reality to answer questions concerning all four types of cause. But this view is also a faith commitment of sorts, for while it may be impossible to prove God's existence, how can anyone prove God's absence? Certainly such an (atheistic) materialist will likely have strong convictions regarding what data would be needed to prove design.

For the record, not all materialists deny God's existence. As we noted, some maintain that matter acts "on its own," but they insist that God made it initially. Thus the outcomes of material causes depend only on the initial state in which God made the matter—plus the possibility of occasional divine input, or reshaping. Such a secondary input to the material system they would term an intervention (or a miracle). Theologically speaking, this view is properly called "deism," or "semideism."[4]

This distinction is important in understanding the recent discussions on "intelligent design."[5] The intelligent-design argument usually focuses on the *visible* actions of God, ignoring the possibility that God can act in an untraceable manner. But in fact, the Bible states that God is acting at every point in all space and time. Thus, "God believers" must decide whether an "evident" act is due to God's desire to show his presence, or because he had no other way to accomplish his purposes.

On the other hand, for those who have committed themselves by faith to nature's autonomy, the idea of intelligent direction of natural causes is simply incomprehensible (even for those who believe in God). For them, a "god" who acts in nature would be the ultimate intruder in a closed system. Such acts thus would necessarily be foreign elements, not part of nature.

As various groups of materialists take up arms on the question of design, they focus on questions concerning how effective independent material mechanisms can be. Are they capable of producing the patterns that we have observed, patterns both in currently living species and in the fossil record? In short, the intelligent-design adherents look for evidence of divine engineering through material inadequacy, while the "matter worshipers" look for material forces capable of producing complicated and often fascinating results without any outside guidance or purpose. Both searches are driven by much the same motive: piety. This is to say that each seeks to honor the object of worship with apologetics and scholarship.

Irreducible Complexity

The primary argument between these two groups has focused on Michael Behe's concept "irreducible complexity."[6] The dichotomy he poses is this: must structures with many complicated, interacting parts be products of design, or could they have developed as a result of impersonal, materialistic forces? First, one must ask whether they came into existence bit by bit or if they appeared suddenly. If they came about in small stages, one must explain their origin and survival through the steps along the way. But if no function is even possible without all the parts present in the first place, then, so goes the argument, it seems that the essential parts must have been there from the beginning.

A partly constructed, nonfunctioning structure would be selected (preserved) only by an intelligent agent working toward a goal.

After all, natural selection, by definition, cannot account for the survival of something that is unable to function. On the other hand, a partly constructed, minimally functioning structure does have immediate value (i.e., it is functional), and therefore it could be selected (i.e., it would survive). A structure with minimal functional complexity would not be "finished" in the sense of perfect efficiency, but it could act as a "functional" starting point for nature to select improvements. Thus in theory it could be "improved" without the aid of some intelligent guiding force operating from outside nature. (Of course, the idea that there is no such force acting, that the natural processes are autonomous, assumes a materialistic worldview.)

An example of such a complex structure is the bacterial flagella, the functional equivalent of a rotating electric propeller. A rotating shaft turns in stationary rings in the cell wall, in turn rotating a long filament with a hook in it that creates the propeller effect. This rotary motor runs on an electric current of hydrogen ions flowing across the bacterial membrane. In his book *Darwin's Black Box,* Michael Behe estimates the minimally essential features to be the proteins for rotor, motor, and paddle. Thus, the first bacterial flagella that functioned at all would have required all three types of proteins—a staggering amount of coordinated information. This is the point, according to Behe, at which God must have inserted a big block of new information to create a "selectable" (functional) flagella.

But must God have done this? Is this the only explanation? In fact, a couple of indications suggest that the bacterial outboard motor could well have been constructed from "spare parts." The bacteria-like cell organelles called "mitochondria" (the cell's powerhouse) have membrane-bound rotating respiratory assemblies (picture tiny turbines) that also run on an electric current of hydrogen ions. They have a very different function, but they do run, rotating, on the same current as does the flagella. Thus they are preadapted to form such a motor. In addition, archeons, the "alternative bacteria" of hot springs and salt lakes, also have a rotating flagellar propeller. However, they have none of the proteins found in the flagella of true bacteria, nor any sign of rotor rings in their membranes.

The point is that there are other ways to rotate flagella, and there are other uses for the parts of the motor. Thus selection for other tasks could have shaped and refined parts of the apparatus before that minimal engine existed. Perhaps the minimal functional state (of the outboard motor) suddenly came into existence through combining existing parts of other devices. And in any case, the argument makes sense only if one assumes the absence of divine guidance.

In short, not all scientists are convinced by Behe's argument that the appearance of this bacterial motor proves intelligent design. We do not know for sure whether bacterial motors did originate, or could have originated, without guidance. But this ignorance does not prove either that God did not do it or that God did do it, nor if God did it, *how* he did it. In fact, it is impossible to find in the study of nature a place where God is forced to do a miracle—that is, a place where supernatural intervention, something that breaks the observable laws and predictable patterns of nature, is the only possibility. After all, how can you stump the omniscient God, especially when presumably he would know the entire universe of biotic probabilities? (That is what "omniscience" means!)

A great deal of the discussion of "intelligent design revealed by irreducible complexity" is based on assumptions that cannot be proven. Since materialists believe in an autonomous universe, they assume that existing forms must be connected by reasonably clear probability pathways. Believers in design, being eager to refute materialism, often assume without evidence that such pathways do not exist. Unfortunately, neither group is omniscient.

Those who believe in God have, I maintain, fallen victim to a false dichotomy when it comes to how God operates in nature.

Many feel forced to choose between "naturalism" (materialism) and "supernaturalism" (God). This choice has been required for many because many people have lost sight of the biblical concept that God operates within nature. I assert again that we should not have to choose between God and so-called natural causes. We can choose *both*.

But the possibility that a mosaic (or gradual scenario) can be envisioned, or even demonstrated, says nothing about probabilities. The cartoons featuring Mr. Magoo, the half-blind blunderer who never gets hurt as he stumbles through chaos, present a story of possibilities versus probabilities: "Oh, Magoo, you've done it again." But the question is whether Magoo actually is doing it himself, or whether he has a guardian angel. Keep that question in mind as we continue through the following chapters.

CHAPTER 7

The Origins of Life

ONE OF THE MOST IMPORTANT QUESTIONS SURROUNDING THE debate over evolution is that of the origins of life. Is it possible for the forces of nature, acting on their own, to turn dead mud into living creatures? This chapter focuses on that question. From a scientific perspective, how did life get started? What mechanisms (causes and processes) were involved? What forces and raw materials were parts of the mix?

Preliminary Considerations

The question of whether nonliving forces could have produced the first living things implies that we know what life is. It also implies that we understand what we mean by "nonliving forces." Again, we need to start with good definitions. For instance, is life best defined in terms of some kind of force, or in terms of some particular arrangement of matter?

This topic also raises theological issues. Some people feel that if any natural mechanisms acted during creation, they would, by definition, be a substitute for God. These people thus maintain that natural mechanisms and creation are mutually exclusive of one another. Others are persuaded that God might actually act through natural mechanisms when he creates. This possibility raises the question of whether the use of natural mechanisms is consistent with the character of a good and righteous God, given that in nature some entities must perish so that others can live and develop.

Basic Characteristics of Life

I acknowledge at the outset that my definition of "life" will upset a materialist. In the chapter on cause, we saw that a central question for designed systems relates to their purpose. As a Christian, I state my conviction that life is indeed created according to a plan, and I am on a quest to determine the shape of that plan. Based on this creation perspective, everything is organized and governed by God's law. Life therefore must be a type of obedience to that law. Defining life is essentially an effort to describe the "shape" of this obedience. To call this shape a form of obedience is, of course, essentially a religious position. But again, to describe this shape is to do science. And whether materialist or theist, biologists are able to recognize the shape.

It may seem to some to be a matter of mere common sense to determine what is or is not alive. But determining what constitutes life is not so simple. A precise definition of life entails some terms that may seem like jargon. Still, each word refers to a unique aspect of what it means to "be alive," and mostly they are commonsense ideas.

As you consider the following terms, remember that the first living thing needed to possess all of them. The terms are homeostasis, boundary, open system, macromolecule, hierarchy, specified complexity, and replication. What do they mean in relation to life?

Homeostasis refers to the idea that all living things engage in metabolism—that is, in life-maintaining activities that could be termed housekeeping. They repair storm damage and sweep out the trash. This requires energy.

Boundary refers to all living things having "insides" that are part of themselves and "outsides" (or environments) that are not. The boundary is the wall, the outer surface between the inside and the outside. It is the site of interaction and exchange.

Open system refers to the flow of things in and out of living things. Open systems absorb needed material and energy and get rid of various wastes. It enables them to build and repair themselves.

Macromolecular means that all living things are made up largely of enormous (macro) molecules constructed from thousands of atoms linked together with absolute precision. You could call these molecules polymers, but unlike industrial polymers such as nylon or Teflon, each macromolecule is unique. These unique macromolecules enable living things to control their own internal chemistries.

Hierarchy refers to the fact that large living things are constructed of smaller parts. Macromolecules are linked to form the (usually microscopic) living entities called cells. Cells in turn act as the building blocks to make tissues and then organs. Organisms are, of course, made of organs (such as kidneys and brains).

Programmed complexity refers to the source of all this. Living things are not complicated collections of random parts. At every level they are the very elegant, precise realizations of blueprints. "Programmed" means that all the processes mentioned above are described/specified in those blueprints. Thus, living things are not simply complex structures; rather, they are specified, described, complex structures, made according to a preexisting plan. (Although materialists do not believe that the plan has a planner, they do acknowledge the plan exists.)

Finally, we all know that living things reproduce or *replicate*. This means that they produce self-realizing blueprints related to, but not identical to, their own. And these blueprints are written upon the chemical DNA.

The significance of all this for the origins of life is that even at its simplest, life is far, far from simple. Living things are self-controlling, programmed. Even at the start, at the birth or origin of any living thing, life requires both a mechanism (defined here as a functioning entity of some kind) and a blueprint, which determines how the mechanism will function. Life's origin requires both to come into existence at the same moment, linked and working. This is not to say that raw materials or spare parts could not accumulate over time in certain unusual environments, but they could not be considered half-alive. Life would have to appear suddenly, in an instant. Life happens when the control circuit between metabolic machine and genetic message closes.

The origin of life is a classic chicken-and-egg quandary. A blueprint is required to produce the mechanism, but the mechanism is required to create the blueprint. For many years some of the most intelligent scientists in the world have been trying to determine how this could have happened, how both mechanism and blueprint could have appeared together and intact. Thus far, no one has been able to do so. From a strictly scientific vantage point, the origin of life remains a grand mystery.[1]

Examining the Science

In considering the origins of life, a brief overview of genetics is helpful. The main distinction within the field of genetics is between the genotype (the gene type or blueprint) and the phenotype (the way the organism appears and acts). One hundred years ago no one had any idea how such a blueprint could be recorded. It was an absolute mystery. Then, about fifty years ago, scientists discovered that the genotype is a set of instructions encoded (written down) upon the molecule DNA, deoxyribonucleic acid, just as a story is written down on a sheet of paper. The instructions describe building a body. To change the body, one must change the DNA.

To understand how DNA changes, we must first understand its structure. A DNA molecule is shaped like a spiral staircase. Each DNA "stair tread" is composed of two smaller molecules called "bases," connected end to end. Each stair tread represents one "letter" of a message on the DNA. There are four bases or letters, thus two pairs. A chemical substance, adenine (A), is paired with the substance thymine (T), and guanine (G) is paired with cytosine (C).

Just as the order of the twenty-six letters of the alphabet determine the meaning of words and sentences in English, so the order of the four bases down the DNA strand determines the genetic message.[2]

What do the bases "mean"? The simplest DNA "words" (codons) are three bases in length, arranged in DNA "sentences" called genes. Each gene describes how to make a particular protein. Proteins are molecular chains (polymers) made up of amino acids. Each codon identifies one of the amino acids in the chain. One could say that it "means" the amino acid, just as the gene "means" the protein.

Here is an example. Alpha hemoglobin is a protein chain with 156 amino acids. The DNA sequence (gene) that describes this protein must therefore have 468 bases (letters) making up 156 codons. Like a story written in English, the DNA language also has "punctuation" codes and "indexing" codes and "synonyms," and more. In addition, each gene sentence is part of a complete DNA "story." DNA stories describe how to make specific organisms. But all these codes are written as sequences of DNA stair treads.

Using DNA to tell the story of life has two important implications. DNA can be copied, and during the copying process DNA (and the message) can be changed. Because the message can be copied, organisms can reproduce themselves, and organisms with many cells (like us) can be constructed—each cell has a copy of the story. DNA can be copied in the following manner. The ladder-like molecule is split down the center and opened up like a zipper. Since the bases in the stair treads are always linked to the same partners, each side of the split ladder can be used as a guide to form a complete ladder (thus C A A T on one side indicates that G T T A should be on the other side). Thus one ladder can be used to make two identical copies, and they in turn to make four copies, and so on. This is called DNA "replication."

With respect to the origin of life question, it is important to remember that the DNA does not copy itself. Rather, it is copied by some of the many proteins that it describes.

Perhaps you can see how difficult it would be to produce the genome (complete DNA blueprint) of the first living thing. Not only would you need the strands of DNA, but also they would already have to have been put together in a sequence that tells a "living" story. But the story is not enough. The proteins to read the story, proteins that the story describes, must already exist. Further, the metabolic machinery of the living thing described would already have to be running under the control of the proteins described by the story. This sounds complicated, and it is. But from this description, unless you are a biologist, you have no idea how complicated all this *really* is. Building the space shuttle pales in comparison.

For this reason, most of the current research into the origin of life is looking for alternative, simpler linked message/machine systems that might coalesce into life spontaneously. The current frontrunner is RNA, a molecule like DNA that can both carry a message and act as a machine. A number of alternative chemical systems, and even inorganic clays, also have been proposed as initial "life carriers." Much has been written proposing and critiquing various ideas, but despite the calm assurances in elementary textbooks, the ability of the forces of nature to produce life from nonlife has not been demonstrated. On the other hand, however, the idea has not been proven utterly impossible.[3]

Rather than describe all of the many proposed scientific theories for the origin of life, I will sketch out the dimensions of the question. The spontaneous appearance of life is rather like the spontaneous appearance of a tornado. Some specific conditions of heat, surface, air movement, and others may lead to a "closed loop" of circulating air that seems to "take on a life of its own."

Could life appear spontaneously in such a fashion? Several difficulties arise. The "living tornado" is not just a physical phenomenon; it requires information. The complexity of the system is specified within the system. This means that a transition must occur from pure complexity of physical state to controlling specifications being encoded. The question of whether unguided nature could make this transition is ultimately a probability question. At this point, no such system transition has ever been described. The level of necessary complexity that would have to build up is simply incalculable. We have no idea. So, then, if there are conditions that could allow the spontaneous appearance of life, what are they? How likely are they? What would be the probability of life appearing if the conditions were met? The truth is that no one really knows. If you strongly desire an autonomous solution, you remain hopeful. If you want to find a gap in causality, you are highly skeptical of all reports of progress.

Could God be involved? Would such a very remote possibility of a "natural" explanation for life's origin make God necessary? Or would it rule God out? Certainly, it is true that for some, establishing *any* possibility, no matter how remote, is enough to rule God out. For others, such a tiny possibility, such a remote chance, is tantamount to being impossible, thus proving God's involvement in the process.

But what is a Christian to think of the idea of chance? This chapter will conclude with a discussion of God versus chance as it relates to the origins of life. But first we will examine whether the DNA message could be changed once it did exist. Such changes could be considered "evolution." They involve processes frequently used to explain the origins of life as well. The possibility of such change also comes down to a question of possibilities.

Changing the Message
Changing the story in an existing DNA message is much easier to understand than the origin of the message. It may happen in several different ways. The way with which we are most familiar is through sexual reproduction. Human beings, like most complex

organisms, represent two versions of the DNA story, one version contributed by each parent. Our bodies therefore are compromise structures worked out by interaction between the two versions. In turn, when we reproduce, what we pass on to each of our offspring is a single "mix-and-match" version of the "human" story selected rather randomly from our two sets. Thus, every offspring is, at least on the surface, a new "roll of the dice," a new and unique recombination of story versions.

Note that with each generation, some of the story variants tend to be lost. I have three sons. None of them got my fingernails. But if thousands of offspring are produced (as in the case of an oyster parent), most of the variants would at some point get passed on. As time passes (especially in organisms, such as human beings, with limited reproduction), the average character of a population is likely to change. This process is termed "genetic drift," and it may indeed be considered "evolution by chance." Note, however, that drift never produces new genes, new story lines. It only produces new combinations of the story lines already in existence.

To get a *new* gene, such as a DNA sequence that will code for a changed protein, one must change the sequence of bases in the DNA. Such a change is called a "mutation." Mutations occur all the time. They may be due to spontaneous errors during replication, or chemical agents or radioactive particles may cause them. Mutations may involve one DNA letter or many, or even whole chunks of chromosomes. They may call for no change in a protein, they may describe a radical change in it, or they may change the amount or the time that the protein is being made. Such changes in a protein "message" (a gene) may or may not have significant effects on the body of the new organism produced.

What, then, is evolutionary change? If the genetic blueprint (genome) is a message written in a language, then evolutionary change must entail changing the letters to change the meaning, and that would change the body. Change in body type (which with fossils is all we have to observe) thus implies a change in the written message (blueprint).

The way that this kind of evolutionary change works can be illustrated by the following game. Try it out for yourself. The goal of the game is to change one English word into another, one letter at a time. Every intermediate word must be in the English dictionary. For instance, "dog" can become "cat" via this path: dog → dot → cot → cat. Each letter change represents a "mutation" in the message. The dictionary requirement represents the "viability" of the message (for example, a change from "dog" to "dag" would not be viable). A path represents a possible process of transformation.

Think of all three-letter sequences of ordinary English letters as a probability universe. Each is either a real word or nonsense. There are 26^3 possible three-letter words (17,576), most of which are not real. The game travels from one real word to another. Of course, a three-letter universe is pretty small, but the numbers go up quickly with more letters. Playing with a five-letter word yields a universe of 11,881,376 (26^5) possible words. Obviously, the game quickly gets harder as the word gets bigger. This is the case for two reasons: the number of possibilities grows astronomically, and the percentage of those sequences that have meaning in English decreases dramatically.

How does this word game correspond to the genetics for evolution? Since every genetic blueprint is written in the same four-letter code, simple logic tells us that all possible DNA blueprints exist in the same probability universe. Like "dog" and "cat," each possible DNA blueprint corresponds to a word in the game just described. And since "being alive" corresponds to "making sense" in the game, only certain blueprints are permissible (can produce a living organism). To be retained (in reproduction), each change in the sequence must yield the blueprint of a viable creature. The goal of the "game of evolution" is to walk from one blueprint to another very different blueprint by changing one base (or a very few bases) at a time, remembering that all intermediate blueprints must "make sense" (be alive). Thus, the genetic equivalent of the word game is the slow accumulation of changes in the sequence of DNA bases leading to transformed blueprints, changed messages, different creatures.

The Evidence in Perspective

The point of this analogy is to state that evolution, defined here in terms of changes in the genetic code that lead to corresponding, different life forms, theoretically *must* be possible. Mutations are inheritable changes in the DNA, producing "new" blueprints. Some will be viable, some will be nonsense. If a path of viable intermediate DNA blueprints exists, then a lineage of descent might follow it to a changed destination, which would be a new form of creature. There are no "intrinsic" barriers, so long as pathways exist. But the existence of pathways in "DNA space" is not automatic. The size of the probability universe (the number of alternative sequences) for bacterial DNA blueprints is around $4^{4,700,000}$. Written out, this figure would be the number one followed by around a million zeros. What human investigator could know the shape of such a probability landscape or predict its pathways? It is indeed possible to propose a "walk" through that space, and thus to picture the "evolution" of bacteria. It is also possible to picture the Duracell rabbit (the renowned battery-powered pink bunny of TV ads) walking across North America. But without a map, you cannot calculate how likely the rabbit is to make it. And can the rabbit walk to Tokyo? Can the rabbit even make it out of town? Who knows?

When we do not know the background probabilities (the map), we cannot predict outcomes. Do evolutionary pathways actually exist? How wide are they? Are there adequate forces in nature to follow them, or would it take guidance? Would evolution require

God in order to work in the way science claims that it has worked? These questions will be addressed in following chapters.

But let us return to the origin-of-life question and apply this same word-game analogy. The big difference between the origin-of-life question and the possibility of evolution is that the latter discussion was based on an analogy with the existing English language. The origin-of-life cannot presume a preexisting language. To put it another way, life is not just a story written in English; it is the equivalent of the English language itself. Explaining mechanisms by which a story might be revised does not explain how language came into existence in the first place. Therefore, evolutionary mechanisms such as those that I have just described could not have produced life. For life to appear by "unaided nature," one must invoke some other natural force.

Creation by Chance?

As we have seen, both the origin of life (especially) and the revision of life seem to depend on highly remote possibilities. In both cases we simply do not know what the background probabilities are. But in any case, would God use chance processes to create? Many unpredictable processes change the genome—recombination, genetic drift, mutation, transposon movement, and more. Are these chance processes undirected? The thesis of Jacques Monod's book *Chance and Necessity* is that pure, undirected, random chance sat at the base of the entire biosphere.[4] This is, in essence, the concept of the "Blind Watchmaker," of a universe that makes itself, having neither plan nor purpose. Does any proposal that relies on chance reject the role of the God of the Scriptures? For many Christians, this is the fundamental reason to oppose evolution.

But to evaluate the strength of this opposition to evolution, we must determine what is meant by "chance." Does it mean that we humans cannot predict outcomes? Or that no one, even someone possessing infinite wisdom, can predict outcomes? Likewise, does it mean that the universe is autonomous, that no one, even some-one with infinite power, can control natural outcomes? These are the assumptions of Monod and Dawkins. The physical universe is their ultimate reality, the object of their faith. They do not simply theorize the existence of a Blind Watchmaker; they worship "him." Given their faith commitment, chance consists of the mysterious edges of "his" ways, the stuff of materialistic miracles.

Perhaps life did originate from nonlife by a very improbable "natural" event. No one knows how improbable such an event would be; we cannot be sure that even a thirteen-billion-year-old universe has enough time for this to be likely to happen. But if life did arise in such a way, by far the most reasonable explanation, probabilistically speaking, is that the situation was set up and controlled by some divine intelligence—the hand of God, if you will. In other words, if such an event occurred, it seems most probable that it happened as a result of divine intelligence, in which case the emergence of life was not "spontaneous" after all. But for those who posit that God does not exist, such probabilities would seem irrelevant.

Ultimately, it is a matter of faith. One must choose: Will I believe in a God who governs nature, or does my faith rest in autonomous nature?

Likewise, one can conclude that a possible mechanism for evolutionary change (according to which various species were not created but rather developed over time, one from the other) does exist. After all, the blueprints of all organisms exist in the same DNA probability space, like squares exist on the same chessboard. Mutation and selection can act as search engines to move a lineage across the genetic "game board." Such movement would be revealed in the different bodies and behaviors of organisms descended from the original ones—in other words, evolutionary change.

However, due to the immense size and unknown landscape of the genetic game board, the question of whether or not such evolutionary trips would need a divine tour guide is clearly a matter of faith. There are, of course, true believers who are sure that they know the answer. Some maintain that such evolutionary trips are impossible, no matter who is the guide. Others are sure that the Neo-Darwinian mechanism—the random search of possibilities— is itself the only guide needed to produce all of life.

But neither of these is the view of a biblical theist. The biblical theist, of course, does not accept the Blind Watchmaker cosmos as the true God. But beyond this, a biblical view of the universe, of nature, of the creation, determines a position on unpredictable events—what we call "chance."

Consider Proverbs 16:33: "The lot is cast into the lap; but the whole disposing thereof is of the LORD" (KJV). The assumption of the writer is that only the Lord God can determine "chance" out-comes. Thus, chance reveals the will of God. Chance is, in fact, the hand of God. This applies to mutation as much as it does to the choosing of a replacement apostle (see Acts 1:26). So a "natural" explanation for the origin of life such as evolutionary processes that unfold through unpredictable events does not exclude the divine. These processes are, instead, processes in which the struc-ture of the universe is contingent on the unpredictable, freewill acts of God. According to this view, God's sovereignty is supreme. This being the case, one could argue that ultimately there is no such thing as chance.[5]

In later chapters we will consider the strength of the scientific evi-dence for just what occurred with respect to evolutionary change. But the central point of resolution is clear. If God rules nature, then nothing that occurs in nature can be considered an alternative to creation. All natural events have a divine dimension. The Bible tells us that God works *through* nature; thus it is wrong to pit God *against* nature.

CHAPTER 8

Darwin's Idea

A RGUABLY, NO EVOLUTIONARY IDEA HAS GENERATED MORE heat than Darwin's central concept, known as "natural selection." The modern theory of evolution (Neo-Darwinism) considers natural selection to be the most important aspect of the theory. And individuals as philosophically diverse as Richard Dawkins (the British atheistic scientist and author of *The Blind Watchmaker*) and Phillip Johnson (the American Christian law professor who has popularized intelligent design) agree that this central idea (which they both dub the "Blind Watchmaker Hypothesis") is intrinsically anti-Christian, that it replaces the loving hand of God with the dead hand of nature. In the preceding chapter we examined this philosophical misunderstanding. But before returning to the theory's implications, let us examine it from a scientific perspective.

The Essence of Natural Selection

Three ideas comprise the essence of natural selection. First is the idea of genetic diversity, which we explored earlier. Organisms of populations differ because of differences in the specific genes that they inherited from their parents. Genes in large part define us physically. No diet and exercise regime could have made me (or, most likely, you) grow to be seven feet tall. But some people are that tall.

The second idea of natural selection is that there is not enough food to go around. Some organisms do better in life than others—they get the food, they get the mate, they dodge the bullet. And these are the ones that produce offspring.

The third idea is that frequently the reason that some organisms fare better is because of their "good" genes, and thus the strong bodies, that they inherited. The strong survive, and they pass on their better genes to the next generation, which will, on average, resemble them. To some extent the poorer genes will have been weeded out. That is natural selection. Up to this point the theory has a strong element of common sense.

In addition to the idea that stronger genes survive and the weaker ones disappear, Neo-Darwinism postulates that this process of natural selection could and did create new forms of life from previous ones. We will examine the evidence for such a possibility in the next chapter. For now, let us focus on natural selection as a mechanism. In other words, let us examine how it works.

Consider the following analogy. I have a large jar in the basement into which I toss extra nails, screws, and bolts. Suppose I want to do a bit of roofing, and I decide to sort out the roofing nails. I reach in, take out handfuls, and toss out everything except the aluminum roofing nails. Gradually, my jar becomes mostly, if not entirely, roofing nails. As it does so, it becomes shiny and half-empty.

Notice the following things about my jar of nails that are true also in natural selection. First, my jar (which represents the available genes of the population gene pool) has become half-empty, just as a population loses genetic diversity with the narrowing of the gene pool. Second, my jar has become brighter as the rusty bolts are removed, just as the population becomes visibly (phenotypically) different—bigger, perhaps. Third, everything that I found in the jar was already there. Likewise, the population has not gotten any new genes through selection, only different arrangements, different percentages of the genes that were already there.

Keep in mind that when all the bolts are gone from my jar, I cannot get them back by discarding the roofing nails. Nor can I create hairpins by sorting the roofing nails. I cannot find anything that was not already there in the jar. And once something is gone from the jar, I cannot get it back.

This does not sound like we are getting anywhere with explaining this theory. Instead of explaining how we could get many different kinds of life, so far it seems that all natural selection is doing is eliminating forms of life by narrowing the gene pool. This is partly because my analogy is too simple. It does not take into account the fact that genes interact. For example, if my roofing nails were made of uranium, and the bolts were made of lead, removing the bolts might produce quite a different and more complicated result. At some point, the jar might blow up the city. By pulling out the lead, I might allow the uranium nails, which were already there, to increase their interaction to a critical level, in theory, creating an atomic bomb. Something new could happen as a result not of addition, but of subtraction.

But did I create something new (the bomb) by selection? Not really. The potential for the bomb was created when I put all those uranium nails into the jar. The results only looked new because I passed a reaction threshold.

This same dynamic can be found in genetic systems. For example, if breeders select (breed) for bigger and bigger paws (or dogs), at some point puppies will appear that have six toes instead of five.[1] The genetic potential for six toes was always there. Breeders "concentrated" this potential (thus increasing the probability for six toes) by dumping the genes for small paws. Again, no new genes were added, even when extra toes appeared. Yet something new appeared.

The Process of Selection
Let us focus on the process of selection. Impersonal nature is neither a carpenter nor a dog breeder who has a reason or purpose for

selecting some nails or canine traits over others. So how could nature "select"? Like a breeder's dogs, some organisms live longer and reproduce more. They may be able to get more food, or to avoid being eaten, or to attract mates, or they may simply be healthier. The aspects of nature that are "selecting" could be, in each case, the nature of the available food, the eye of the predator, the desire of the mate, and the level of genomic integration. The results might be thicker beaks, shadowed colors, a long tail, or higher energy levels.[2]

There is quite a bit of evidence that this process does indeed cause change in nature. Natural populations have been observed as they respond to environmental crises with selected changes—such as an increase in the thickness of finches' beaks in response to drought, pollution resistance in weeds, or bacterial resistance to antibiotics. Of course, such changes do not constitute the production of major new body types, but that is hardly surprising in the few short years we have been looking for it. In fact, the degree of change that can be produced in the lab "in a few short years" can be quite amazing: the average values for some body characteristics (the number of bristles on a fruit fly, for instance) have been changed by ten standard deviations.[3] This observed rate of change is far higher than the rate seems to be for even rapidly changing fossil lineages. The better question may be this: What makes populations so stable most of the time?

We see that apparently in most populations the raw material for selection—the available genetic diversity—is quite high. But remember, selection removes diversity. So biological change cannot continue indefinitely by selection alone, otherwise the diversity eventually would disappear completely. Thus, as a mechanism for producing new creatures, natural selection is "natural" and powerful, but incomplete. As a builder, it can lay bricks, but it cannot make new ones.

Mutation: Creation by Chance?

Now we turn our attention to the rest of the accepted evolutionary mechanism: the action of mutation to make the "new bricks" (new genes) required by that selection. Again, natural selection certainly cannot make evolution happen all by itself. It can remove the bolts from our hypothetical jar of nails, but it cannot add new nails. It needs a nail-maker! This fundamental limitation was why Darwin's theory was not widely accepted until around 1940. It seemed clear to many that even if selection could give direction to evolutionary change, it could not be the whole story. Selection added nothing new to the gene pool; its products were merely new combinations of existing variants. No matter how powerful a tool selection was for exploring existing potentials, eventually it would run out of gas. It was inherently incapable of producing new forms of life. What was needed was a source of new variation, new alleles (alternate forms of genes).

That new source was mutation, defined as unpredictable changes in the genes. Mutation as the source of new variation was combined with the sorting effects of selection to form the "modern synthesis" of evolutionary theory, termed Neo-Darwinism.

Let us now examine the efficacy and implications of mutation. As we noted, a mutation is an unpredictable change in the genome (genetic code) of a living thing. Such a change can happen spontaneously. That is, within the DNA itself there exists the potential for internal "accidental" (seemingly arbitrary) change. Mutations can also result from outside forces such as chemicals, radiation, or biological agents. The extent of the change can range from the replacement of a single base in a DNA sequence to the loss of entire chromosome arms.

That such changes can and do occur is indisputable. Our understanding is rooted in the same kind of scientific discovery that serves as the basis for DNA forensics commonly used, for example, in police investigations.

As far as we can tell, there is no way to predict which specific changes will result when a mutation occurs. From all appearances, there is no relationship between a mutation and an organism's need for something better in its genetic makeup. Various questions arise: How likely are useful mutations? Are not mutations usually deadly? Do they not appear by "chance"? If so, how might such a chance process be related to creation?

The Effects of Mutations

To understand the effects of mutations, it is helpful to understand the function of proteins. Proteins, the chemical "workhorses" of the cell, are constructed as subunits called amino acids form long chains, which fold into specific shapes. The specific order of the amino acids determines the function of the protein, and genes determine the specific sequences. If the base sequence of the DNA is changed, then the protein description it encodes will also be changed.

Mutational effects may range from no effect at all to a complete loss of function. There is no effect if the mutant (changed) codon (DNA codeword) has the same meaning as the original codon. The message and the protein are unchanged. But a mutated codon is more likely to call for a different amino acid, and the result is change. Usually the "new" amino acid called for is chemically related to the original one, and so the effect of the mutation is relatively minimal, even undetectable. Since their meanings are intact, such mutations are not harmful.

Sometimes, however, the mutation calls for an amino acid of a different sort. In that case, the protein might be inactivated or possibly have a changed function. For instance, the mutation that causes sickle-cell anemia changes a single amino acid to a different type. The changed protein clumps together under certain conditions (low oxygen), leading to all the symptoms of the disease as well as to defense against malaria.

In contrast, a large number of hemoglobin variants (resulting from mutations) have no easily detectable or unusual properties. The common view among scientists is that most "favorable" mutants are subtle variants of this sort. That is, they survive the process of natural selection, but the effects of the mutations are minor.

Proteins play many roles. They catalyze chemical reactions, they ferry things across membranes, and they produce physical movement. But the most dramatic effects occur if there is a mutation in one of the genes that affects communication with other genes. The best-known examples are the Metazoan Hox genes. The Hox genes are sets of genes that activate entire genetic programs, causing them to speed up, slow down, turn on, or turn off.[4] Might this be the means through which to achieve an easy creation of new forms?

Here is a specific example that illustrates how powerful, and how limited, such mutations would be. Antennapedia is a Hox gene that produces a protein that instructs fruit fly tissues to make a leg. A mutant form of the gene makes this protein in the head, causing the fly's antenna cells to grow into legs (-pedia). But that does not mean that antennapedia is a "magic gene," able to snap complete legs into existence.

A gene homologous to antennapedia exists in mammals ("homologous" means that the two genes have almost the same DNA sequence and describe proteins that are almost identical). If the gene is placed into the fruit fly genome (by genetic engineering), the mammalian gene can also trigger the conversion of antennae to legs. But the mammalian gene calls forth fruit fly legs in the fruit fly. In mammals the same gene evokes completely different structures. So then, a Hox gene mutant may indeed cause a major morphological shift, but usually it is one based on its normal role in determining the standard morphology of its lineage. Thus, mutating Hox genes do not create new genetic programs. The most they can do is switch on a different existing program.[5]

What is the significance of all this? What can we conclude with regard to mutation, this "creator by chance"? Well, in the first place, a mutation does not really create; rather, it discovers new things in the "mine" of the DNA probability space that were there in the first place. Think of the analogy of a miner panning for gold. The evolutionary mechanism is akin to two partners. Joe Mutation scoops up shovels full of gravel, while Nat Selection washes them in his pans, looking for nuggets, dumping the rock, assuring that only the gold will remain.

Getting rich (evolving) will depend, of course, on the industry of our miners, but it will depend even more on how good their mining claim is. They need to be working in a fruitful section of the stream. If they do "strike it rich," we can ask how they came upon this particular section of stream. Was it sheer luck? Or did someone point the way?

Let us consider for a moment the size of the "genetic mountain range," starting with a very small part of the mountain range. Consider "one protein space." Proteins are made of 20 different kinds of amino acids, so a protein space equals 20^N (N being the length of the protein). Thus if a protein were just five amino acids long, there would be 3,200,000 possibilities. (Proteins, however, have an average length of around 100 amino acids.) It would take 15 DNA bases to code for our tiny 5-amino-acid "protein." There are 1,073,741,824 possible DNA sequences of that length. That sounds like a very, very large mountain to search. But it is less than a grain of sand compared to the whole DNA range. And even so, only 45 of the local possibilities of our 5-amino-acid protein mountain can be reached by changing just one base. In other words, our miners are not searching the whole mountain range. They are searching their local neighborhood, and they are very, very near-sighted.

The Evidence in Perspective

What does this mean? It means that someone who insists that evolution's success is due to unguided happenstance must take on faith either that the entire mountain range is filled with gold or that life just happened to find some very rare rich spot. There is no way to prove either proposition. Someone else might assume that God carefully created the mountain range—the probability universe that life is exploring—with paths that life could follow to develop autonomously. Or, alternatively, one might assume that God guides lineages down specific paths to reach predestined endpoints. To detect the difference, you would have to know the geography of the mountain range (a genomic universe). And of course, only an infinite God could know that.

Although we cannot predict what new versions of genes will mean, they do provide new diversity for the gene pool. Organisms with these mutant sequences must also try to survive, grow, and mate. If they leave more offspring than organisms with the original sequences, they are being selected *for*, and the "new" (mutant) gene (allele) is being established in the population; if they leave fewer, they are being selected *against*, and the "new" form probably will be lost. The standard Neo-Darwinian idea has been that mutant genes that are slightly better adapted to survival in the environment will slowly displace the original sequences. When enough small-effect mutants accumulate, they add up to a major morphological change.

It seems clear that this mechanism does have the potential to search for and find genomes for viable organisms in genetic probability space, provided the genomes are really there and are connected by viable paths of intermediate genomes. But we probably will never know even the local shape of genetic probability space well enough to know whether the transitions in the fossil record were likely or unlikely. Without knowing those background probabilities, we cannot say whether this search was likely to have been guided, or whether it happened by "chance," except as we respond through faith.

Natural selection certainly is a powerful natural mechanism. So are tornados. Does natural selection have any wider meaning than a tornado? Whether impotent or all-powerful, is selection indeed a "Blind Watchmaker"? First, keep in mind that "natural selection" is just a metaphor based on selections made by intelligent people— nails from a jar, or paw size in a dog population, or best-of-show in an art contest. But the thing that "selects" in natural selection is nature itself. More specifically, the selection is being done by the circumstances of a specific situation in space and time. It is carried out by a population's local environment and genome. Obviously, what this means to you depends on your view of nature.

Is natural selection a "purposeless material process," as Phillip Johnson and Richard Dawkins would have it? Obviously it involves *natural* processes, things that happen in nature. But the Bible views none of these processes as purposeless or autonomous. They are, in fact, the very "natural" events that are said to show God's providential hand at work—getting food, reproduction, life, and death (see Psalm 104:27-30, for instance).

So what is the source of the idea that natural events are "purposeless"? This is not science, but rather a nonbiblical philosophy of nature. Dawkins must first assume what he says he is trying to prove: nature has no purpose behind it. Johnson probably does not really think that, but since he is arguing with Dawkins's crowd, he speaks the same language on this point. Or perhaps Johnson really does not consider everyday events, such as the deaths of sparrows, to be directed by God. Perhaps he retains the idea of God's purpose for the overall process but not for the details. (It is hard to be sure what he thinks on this point.)

Darwin's metaphor gained much of its ideological power from the fact that it *was* a metaphor. Natural selection seemed to place in the hands of blind nature an ability previously considered to be a unique aspect of intelligence: choice. Three hundred years earlier, in the heyday of Thomas Aquinas's synthesis of Aristotle and Scripture, natural selection would not have been viewed as problematic for theology. Thomistic nature was not blind; rather, it was endowed by its Creator with powers to produce forms. Natural selection would simply have been viewed as a theory of how nature went about obeying its creator. However, when the Enlightenment postulated an independent nature as a theater for autonomous human dominion, the Blind Watchmaker followed naturally. Given those assumptions, Darwin's metaphor replaced intelligence with blind mechanism.

But can anyone be sure that there is no intelligence involved in the daily events of nature? We should reconsider the nature of God. A scriptural view of nature assumes that the purposes, plan, and hand of God are behind each situation in space and time. Thus "natural" selection would also always actually be "supernatural selection." It is true that a materialist's view of nature assumes that natural situations are happenstance, without purpose or plan, and certainly not governed or guided by the divine hand. But then, the materialist's view of natural selection might be better termed "materialistic selection."

The materialist view of selection makes sense if one supposes the existence of a very elaborate and complex pool of genetic potential. But then the question arises as to where this original genetic potential could have come from. Would not the preexistence of immense genetic possibilities be an argument for a creating God? On the other hand, if one does not suppose the preexistence of that high genetic potential, and yet the evidence demonstrates that new creatures have been produced through natural selection, then an equally good argument could be made for active divine guidance.

Thus when selection produces change, a scriptural view sees it as due to God's presence, while a materialist view sees it as God's absence. But such evidence cannot prove *either* view, for proof would require knowledge of the almost infinite "mountain range" of genetic probabilities. Individual people see things in light of, and thus as evidence for, the views to which they have committed themselves.

I offer here a double quotation from a book opposing "Darwinism."[6] The nineteenth-century theologian Charles Hodge quoted botanist Asa Gray thus: "If Mr. Darwin believes that the events which he supposes to have occurred and the results we behold around us were undirected and undesigned; or if the physicist believes that the natural forces to which he refers phenomena are uncaused and undirected, no argument is needed to show such belief is atheistic." Since Darwin did view natural forces as undirected and undesigned, Hodge concluded that Darwin's theory was indeed atheistic. Still, as Hodge concluded, "That does not mean, as said before, that Mr. Darwin himself and all who adopt his views are atheists; but it means that his theory is atheistic." As a matter of fact, Darwin was not an atheist.[7] But Darwin's theory Hodge considered atheistic because it viewed natural forces as autonomous and undirected. By this measure, Hodge would call atheistic the views of anyone today who understands natural forces to be autonomous—uncaused and undirected—whether it be Richard Dawkins or Phillip Johnson.

Both selection and mutation raise theological issues, especially as applied to creation. There are those who never tire of pointing out that using selection to produce new forms involves the death of countless organisms. This, they argue, is beneath the dignity and goodness of the God in whom we profess to believe. But remember that the forces of selection, the control of births and deaths, are precisely those that the psalmist uses to pinpoint the sovereignty and authority of God over nature in Psalm 104. Remember that *all* things die, the "fit" as surely as the "unfit." If God determines the time of death even of sparrows, then all that (super)natural selection means is that God is always acting with purpose in nature.

Perhaps our problem is that we are reluctant to admit that God has the right to bring death to his creatures. Or perhaps we do not understand what an infinite intelligence would mean when we discuss "teleology." B. B. Warfield pointed out a century ago that Darwin's understanding of teleology was defective.[8] In God's universe peacocks do not exist merely to give eagles a meal. Nor do they exist to allow some evolutionary descendant species to be shaped. They exist because God likes peacocks (Job 39:13). In the process eagles get fed (Job 39:29-30), and new species are born, but the primary purpose of peacocks is to exist to the glory of God. An infinite God holds together all purposes simultaneously without tension.

CHAPTER 9

The Origins of Species

IN THE PRECEDING CHAPTER WE SAW HOW CHANGES TO THE GENETIC code can change the nature of a living thing. But how radical can this change be? It is one thing to consider one kind of cat changing over time to become another kind of cat (or cats). But can natural selection account for the emergence of very different living things?

Among the fundamental ideas associated with evolution is the concept that new species arose from previously existing ones. In fact, according to evolutionary theory, all of life—bacteria, roses, pine trees, squirrels, cows, and people—developed from a single source. Is this possible? And can it be consistent with what the Bible teaches? We will examine this aspect of evolution in a later chapter, but for now let us focus on the origins of species.

To determine whether one species can evolve from another, we first need to define the term "species." Unfortunately, the concept of species is not clear-cut. Part of the confusion arises from history. The term, which is rooted in Latin, was used first during the medieval period, during which identification and classification were dominated by Platonism. "Species" referred to metaphysically distinct groups shaped by "specific" immaterial essences. According to the Christian spin on this Greek idea, essences/archetypes were created by God, the divine architect. Such essences were thought to determine the characteristics of everything from metals and rocks to fish and birds. Objects were alike because they shared the same essence or archetype. Museums were filled with "type" specimens, those that were "typical of," or close to, that ideal form.

The obvious way to identify a species was to compare structures inside and out. This "morphological" idea could be expanded to include everything from genetic sequences to behaviors. In theory, morphological criteria should make it easy to identify species. But this method has its shortcomings. For instance, many seemingly identical groups of biological creatures (such as the western pocket gopher) are divided into numerous cryptic "sibling" species. These different kinds never interbreed, even though they look almost exactly alike (at least to us). Should each of these "hidden" breeding groups be considered a separate species?

On the other hand, other species are composed of interbreeding individuals that vary wildly from one end of the range to the other. Still other groups (or species) do not interbreed, but could if they chose to. In fact, they will interbreed in the lab or at the zoo. Which ones should actually be considered species? There is no simple answer; it depends on the chosen criteria.[1]

The "biological species" concept identifies a species as a group of organisms that breed together in the wild. But even this has difficulties. For example, it is not possible, based on this concept, to identify species from fossil records, since there is no way to observe the breeding patterns of fossils. Beyond this, there are "parthenogenic species," species that never breed, but produce offspring in some way other than sexual reproduction. Then there are stable "wild species" that do some interbreeding with other stable species (a group of such interbreeding species is termed a "syngamneon"). Should they be considered one species instead of two?

Examining the Evidence

The questions for this chapter are these: How do species (whatever they are) arise? Did they appear out of nowhere? Or do they arise from an existing species? If so, how? It is not easy to reach a conclusion through observation. Current thinking holds that species usually exist for a long time, perhaps millions of years. A new one is rather unlikely to pop into existence in front of your camera. But in some cases, it comes close. I will mention two brief examples, and one longer one, and then draw some conclusions.

The first example is the apple maggot fly, *Rhagoletes*. The flies do their "nuptial dance," mate, and lay their eggs on apples. Three hundred years ago there were no apple maggot flies. There were no apples in North America. There was, however, a hawthorn maggot fly. Sometime after the introduction of apples from Europe, a few hawthorn flies switched over to apples. The apple flies are not very different from their hawthorn-loving ancestors, but they mate and lay eggs at different times. The two groups do not mate in the wild. They now act as two distinct species. Admittedly, that is not very impressive. Nothing is new except food and location. But it is very recent, perhaps a species formation in process.[2]

Another example of speciation in process is the Plethodontid salamander *Ensatina eschscholtzii*.[3] *Ensatina* lives in the mountains around the San Joaquin valley of California. It has a sort of "hairpin" distribution. It you start south of the valley and follow the western range, these salamanders breed with their neighbors, which differ in color and habits. This breeding pattern continues across the north end, and south down the eastern range. When you get back to the point where you started, south of the valley, you will be looking at a different salamander. The populations at the two ends of the hairpin will not breed with each other. They appear to be different species, but only there in the south. Are they one species or two? Or are they one becoming two? It depends on the chosen criteria.

My third example, the Kodiak bear, is a bit more colorful than fruit flies and salamanders. Actually, this example involves both the Kodiak bear and the polar bear. The Kodiak is a subspecies (i.e., part of a larger species) of the brown bear species, *Ursus arctos,* that ranges across North America, Europe, and Asia. The brown bears of Europe, Russia, and the Kodiak Islands are all part of the same interbreeding species. Weighing over 1,600 pounds, Kodiak bears are the largest land carnivores. But like most bear species, the brown bears are omnivorous, with grinding molars able to process rough food. The DNA "clocks" of these bears indicate that the first brown bears lived more than 850,000 years ago.

The polar bear, *Ursus maritimus,* can be almost as big as a Kodiak bear. It differs from most other bear species by virtue of a number of unique features that adapt it to a semiaquatic, arctic way of life. It is almost entirely carnivorous, living mostly on seals. Its molar teeth are more like a cat's carnassials, scissor-like teeth to cut meat off the bone. Its claws are relatively short, curved, and sharp—again, more like a cat's. The skull is narrower, with greater bite strength than that of the brown bear. Its elongated legs, neck, and body allow a faster running speed, with dashes up to twenty-five miles per hour.

The polar bear is further characterized by a series of unique arctic adaptations. The white fur is not just camouflage; it is actually transparent. The hairs collect ultraviolet light (a sort of fiber-optic effect) and transmit it to the skin surface, which is actually black. This absorption is so efficient that a polar bear is visible on UV film as a dark blot! The fur is also such an effective insulator that these bears cannot be spotted with infrared (heat-sensitive) cameras. Another "arctic special" is the surface of their paws, covered with short hairs and with pits, providing a sort of suction-cup effect to avoid slippage on the ice.

The polar bear is also much more aquatic than any other bear. It will swim for hundreds of miles in the arctic seas, up to eight miles per hour, even swimming underwater. The forepaws are especially large and partly webbed for swimming. The pelvis is modified to allow the hind legs to trail behind as a rudder. The body is covered with a three-inch layer of blubber (typical of aquatic mammals), except for the inner sides of the hind legs, which are held together in swimming. The fur has heavy oil content, allowing water removal, and the hair is hollow for added buoyancy. Finally, the eye is protected underwater by a "nictitating membrane" (third eyelid). Polar bears are indeed remarkable creations!

But the key point is that the genetics of the polar bear indicate that it is actually a very young species. Polar bears appeared during the Pleistocene, perhaps as recently as two hundred thousand years ago. So where did this unique bear come from? Did it have ancestors, or was it directly created? The genetic evidence indicates that polar bears are very closely related to the brown bears of the Kodiak Islands, *Ursus arctos*.[4] In fact, the brown bears of the Kodiak Islands are more closely related genetically to Polar bears than they are to other members of their own brown bear species. Polar bears and brown bears will actually breed in a zoo and produce fertile offspring.

What does this mean? Put it this way: if the Kodiak and polar bears are viewed as first cousins, the other brown bears of Alaska are their second cousins, and the brown bears of Russia are their third cousins. Not so long ago a local population of brown bears split. One half became the polar bear species, and the other half remained as part of the brown bear species. A new and different species was born. So species formation, like birth, is an event that occurs at a specific place and time. "Infant species" have ancestors and relatives.

Remember that significant genetic, morphological, and behavioral changes happened during the formation of the Polar bear species. Polar bears are obviously very different animals from their brown bear ancestors/cousins. Where did these changes come from? We will consider this question in a later chapter. For now, simply note that it did indeed happen.

Some would argue that there could have been a direct creation of polar bears. God could do that, of course. But if so, then God created a "new" species with genetics that look almost exactly like that of a single isolated population of a previously existing species. If this were a case of "special creation," we would expect the polar bear's genetics to be at least at equal distance from the whole brown bear species instead of being so closely related to the Kodiak bear. Also, the location of the Kodiak Islands is exactly where we would expect a few bears to get isolated and develop a taste for deep winter—on the Bering Strait during a cold part of the Ice Age. Of course, God can do anything. God could have created the polar bear with the appearance of having developed from the Kodiak bear. But if God writes fictional narratives into the creation to throw us off, how can we trust either nature or the Word?

Have we just shown that God was unnecessary for the creation of polar bears, or even for salamanders and fruit flies? Of course not! Not if one holds a scriptural view of the relationship between God and nature. Natural events are the ways in which God accomplishes his purposes in nature. If God weaves the baby in the mother's womb, surely he can weave a species in nature's womb. If the second weaving is beyond God, so is the first. If the first is God's work, so is the second.

Let us tie this back into the discussion of the four kinds of cause from chapter 6. For God to make polar bears from brown bears does not mean that God's final cause—God's intention to make the great bear of the northern ice—was not realized. This bear is what God intended. Nor does it mean that the divine formal cause—the polar bear blueprint—was compromised. The polar bear has the characteristics that God intended it to have. Whatever efficient cause (process) God employed to make the polar bear, the Creator's hands were upon the process, guiding and creating. It does mean that God chose to use brown bears as the raw material (material cause) as opposed to polar ice, but surely the potter is free to use the sort of clay deemed appropriate for the task at hand. And usually a potter's choice of clay reflects the goal of the proposed project. Is this not the way in which a skilled human crafter proceeds? If *you* wanted to make a polar bear, what would you use?

Biological Diversity and the Genetics of Populations

The term "evolution" refers not to individual organisms, but to populations of organisms. Individual organisms do not evolve. *Groups* of organisms—populations—do evolve. Individual organisms are never exactly alike, and the genetic diversity that makes them different is the raw material for change.

This biological diversity was first viewed as a defect, an imperfection of creation, not as a created gift. Taxonomy, the science of classification, was originally embedded in Platonic categories. According to this system, a "species" was considered a unique entity, but each individual member of the species was a slightly defective "offprint" from its eternal ideal form. This is the reason that early taxonomy looked for "type" forms, exemplary organisms that would be as close as possible to their eternal "ideal types." Under this system, one can see why population diversity was considered an interference with good science.

But diversity is a fundamental characteristic of populations. Each organism has its own genes and blueprints. Usually each organism has two slightly different copies, one from each parent (a condition called "diploidy"). It follows that every organism—and every population, of course—has built-in diversity.

This difference in the genetic message is considered to be the key to understanding how populations change. Alternate forms of the same genes ("alleles") can have profound effects. For example, sickle-cell hemoglobin is produced by an alternative form of the gene that makes (normal) beta hemoglobin. An individual can have two genes for normal hemoglobin, two genes for sickle hemoglobin, or one gene for each. If both genes are sickle-cell, then the person has the disease sickle-cell anemia. With two normal genes, there is no anemia, but the person tends to become gravely ill if infected with malaria. But if one copy is sickle and one is normal, the person does not have anemia and is also resistant to malaria. In this case, life and death can depend on which forms of the genes mom and dad had, and which ones they passed on to their child.

Evolution has been defined (by R. A. Fisher) as simply a change in a population's gene frequency (or gene pool), a genetic change that causes a change in the average characteristics of a population.[5] As we noted earlier, such "evolution" does not produce new genes, but it can produce new genetic combinations. And in certain locations, such as in populations "on the fringe" due to poor-quality environments, those new combinations may enable survival—and permanent change.

But organisms within a population cannot be too different, for they must breed with one another. Each organism and its mate must share common genetic blueprints that will successfully work together to produce viable and fertile offspring. Think of the effect of mating a horse and a donkey. The result, a mule, certainly is a viable animal, but one that is almost always sterile. Population diversity has necessary limits.

When it comes to evaluating evolution as a mechanism for change, the question is whether and how new genetic combinations can fuel permanent and significant change in populations. How can a population's gene pool change? What would be the source of new possibilities? Clearly, natural populations have within themselves the genetic potentials to produce new genetic combinations, new species. But that does not mean that nature can create without need for the hand of God. Rather, this strong genetic evidence shows how God chooses to work in nature to develop his creation. For the believer, such an unfolding of nature's potentials is simply the clear evidence of the hand of God. But how effective is it? The question for the next chapter is whether the process of forming new species is enough to explain the drastic transformation of a lineage. Did the first whales walk on dry land?

CHAPTER 10

The Question of "Missing Links"

THE QUESTION OF "MISSING LINKS" IS RAISED IN ANY DISCUSSION of evolution in regard to relationships between and among living things. This is a very legitimate argument against "macroevolution," the idea that species evolved from other species. If one says that today's creatures are descended from very different ancestors, obviously there should be evidence of in-between types.

In this chapter we will look at the evidence related to missing links. First, however, we need to make a couple of clarifications. The first is a question of logic: Is missing a link clear evidence against the idea that some species descended from another? Not necessarily. Even if all forms of life are related, fossil evidence for *every* lineage could not possibly be available. For one thing, only a tiny fraction of all the things that have been alive have been preserved. Many will never be found. The prediction of common descent would be that there will be at least *some* significant evidence that intermediate forms existed. On the other hand, the "special creation" of species (or kinds) predicts a complete absence of all intermediates between created types—that is, the missing of *all* links. So if even a few intermediate types can be found, this would be enough to call into question the idea of special creation.

The second matter for clarification relates to the difficulty of identifying a "type" (created category). This concept presumes the existence of a type form, a created category that sets limits on the biological nature of its type. Those who advocate for the theory of special creationism hold that God created a biological world divided into such types. They term each type a "baramin," a word coined based on the Hebrew words for "creation" (*bara*) and "kind" (*min*).[1] Since baramin would be defined in terms of a certain set of unique features, special creation implies that creatures should not exist that have only some of a baramin's defining characteristics or that display a mixture of the features of two types. But deciding what that means is not easy.

For instance, will a baramin be defined in terms of a species, such as the white-throated sparrow? Or will it be defined on a more inclusive level, such as parrot, bird, or even vertebrate? And how can we identify a legitimate missing link if when such a link is apparently found, it can be disregarded simply by redefining the type? Obviously, one cannot make a case for missing links if the baramin in question can, without limit, be broadened and redefined to include them. Furthermore, a special creationist can decide to call the missing link a new, unique, and separate baramin. After all, for all we can tell for sure, every fossil ever found could be from a separate, nonbreeding type.

Here is an example of the problem. Let us define a mammal (as Linneaus did) as an animal with hair and mammary glands that gives birth to live offspring and that lacks a cloaca. Based on this definition, the platypus appears to be a "not-missing" link between mammals and reptiles. It has hair and gives milk, but it lacks nipples, and, like a reptile, has a cloaca and lays eggs. Further, its scapula (shoulder blade) is reptilian in form, not mammalian. Now obviously, the platypus is not a semireptilian ancestor of modern mammals. It is alive now. But the categories "reptile" and "mammal" apparently are not as isolated as we thought. Instead of calling the platypus an intermediate link, one can simply broaden the category "mammal" to include it. Or create a new baramin separate from both placental mammals and reptiles. The point here is that for those committed to the conviction that no intermediate links exist, there is no way to prove them wrong.

Bird or *T. Rex*?

The platypus is still alive, so it is no one's ancestor. But how about a fossil record that might point toward a distant "grandparent" of one of today's forms? The obvious choice is *Archaeopteryx lithographica*, the Jurassic "early bird" first found in the Solnhofen limestone in 1861. The chicken-sized *Archaeopteryx* has been claimed as both bird and reptile (as a theropod dinosaur). It has been everyone's favorite missing link, from Darwin's day until now, because of its mixture of avian and reptilian characteristics. Is it just an aberrant bird, a sort of avian platypus? Or is it an intermediate link that is not missing? Baraminic typology would expect a fairly clean distinction; evolution would look for mixed characteristics.

How much of a reptile was *Archaeopteryx?* Wellnhofer (1990) listed eight typical reptilian characteristics: (1) long spinal tail, (2) saurian pelvis, (3) vertical femur, (4) hand with clawed fingers, (5) gastral ribs, (6) ribs without uncinate processes, (7) reptilian growth pattern, and (8) reptilian lung type.[2]

In addition to displaying all of these traits, *Archaeopteryx* had teeth rather than a beak (although teeth are found also in some birds of the late Cretaceous). Likewise, clawed hands (wings) are found on the chick of one modern bird, the Hoatzin. However, the balance of the comparison so far would be in favor of calling the fossil a reptile.

"Nevertheless," Wellnhofer continued, "the perfectly developed plumage of *Archaeopteryx* makes it certain that the animal did fly. *Archaeopteryx* represents an advanced stage in the evolution of flight. Its main feathers show the asymmetric, aerodynamic form typical of modern birds." So, it has flight feathers—bird wings—

but how well did it fly? With no sternum and a small wishbone, there was very little surface for the attachment of the pectoral muscles. The ulna showed no sign that the major wing feathers were secured by ligaments, and the third digit of the manus (which generates the power stroke in modern birds) had relatively small feathers. Wellnhofer concluded that the animal had a lifestyle rather like a flying squirrel: a good climber that coasted to the ground and then ran for its life to the next tree. It flew, but not well.

Archaeopteryx clearly was an intermediate between birds and reptiles. Is it a missing link? That depends on the definitions of kinds, of course. If "bird-kind" is broadened to include everything with feathers and flight, it is a bird. If "theropod dinosaur" is broadened to include the possibility of feathers, *Archaeopteryx* is a dinosaur. Or one can assign it a baramin of its own.

Was *Archaeopteryx* a fluke? The isolated exception that proves the rule? Well, exceptions do not prove rules, but it is true that for a long time *Archaeopteryx* stood in lonely glory. In the last couple decades, however, that isolation has changed dramatically. A wealth of new specimens from China has joined *Archaeopteryx* on the bridge from reptile to bird.[3] They stand on both sides of the gap. Feathers of some sort, as well as a broad sternum and uncinate rib processes, have been found on six different sorts of small theropod dinosaurs. In *Caudipteryx* there are true vaned feathers arranged into groups (primaries, secondaries, rectrices) typical of flight. However, the skeleton has no fight capability. Down feathers have been found on specimens less like birds than the Tyrannosaurids, suggesting that *T. rex* might have had (big) downy chicks! The skeleton of the oldest Dromseosaur (the "raptors" made famous by the film *Jurassic Park*) not only has down, but also has a skeletal structure closer to the birds than any other dinosaur.

On the bird side of the gap a whole series of such links has begun to turn up. *Rahona ostromi*, a fossil bird close to *Archaeopteryx*, had a killing claw on its foot like the Dromseosaurs (raptors). In fact, *Archaeopteryx* also had such a small claw. Next, thousands of specimens of an early a crow-sized bird called *Confuuciusornis* have been found in sediments of the early Cretaceous. They show a deep thorax with a strut-shaped coracoid (a bone supporting the shoulder) and a longish pygostyle (fused tail vertebrae).[4]

Following that, there is the "opposite bird" (enantiornithine) radiation. (In this context "radiation" might be defined as several similar groups of creatures developing from a common ancestor.) Opposite birds represent a wide variety of birds with alular feathers for slow flight control, rectriceal tail fan for maneuvering and braking, a wing-folding mechanism, and an opposable hallux (toe) for perching. But the opposite birds also had wing claws, a dinosaurian "footed" pubis, and teeth. The enantiornithines dominated the skies throughout the Cretaceous until they were replaced by the radiation of the euornithes, or true (modern) birds. Euornithes have an elastic furcula (wishbone) and deep sternal keel, allowing much larger flight muscles. They lack teeth and wing claws (almost always), and have a fused synsacrum. (In terms of a modern chicken, the synsacrum would be the back.)

I have thrown a lot of terms and fossils at the reader here. I know this can be tedious. But I want to show the scope, complexity, and depth of the evidence for just one lineage. Are there missing links in this story? One can, of course, divide it into many segments to retain the idea of independently created types. Each segment could be a created type. But if so, God made them in a series that in morphology and time suggest continuous development. To me, this is akin to believing that God created a young earth with the appearance of age. It is fictional history.

It seems more reasonable to assume that such fossil data are intended to give us a clear picture of how creation actually took place within a developing lineage. From dinosaur to modern robin, the pieces are in place. Further, as time passes, more pieces are being found in other major transitions, such as fish to amphibian, reptile to mammal, and land dweller to whale. And almost week-ly it seems that we hear of a new hominid fossil, a matter that we will pick up in chapter 13.

Punctuated Equilibrium

Can the evolutionary mechanisms that we examined in the pre-ceding chapter account for such transitions? What is involved in a major transition of a living entity's form? I have presented evidence that new species formed and that certain fossil lineages seem to have changed dramatically through time. Nevertheless, the reader may feel that I am ignoring evidence, that evolution is a "theory in crisis." Specifically, I have not discussed "punctuated equilibrium" (or "punk eek"), and certainly it is true that a great deal of the con-troversy in evolutionary thought over the last few years has cen-tered around this idea from Stephen Gould and Niles Eldredge.[5] These two paleontologists state that the fossil record shows that evolutionary change either happens rapidly or does not happen at all, that evolution proceeds by fits and starts.

But first, let us review the predictions of both Neo-Darwinism and Special Creationism. Neo-Darwinism holds that useful new mutations will have only minor effects. Thus, major change occurs as these mutants are gradually accumulated by natural selection. Therefore the fossil record should show continuous, slow, gradual change. Major transitions will take a long time to occur, and there should be vast numbers of intermediate forms.

On the other hand, Special Creationism (old-earth version) holds that living things were created as "baramin," limited biotic groups. Although mutation and selection, and even new species formation, can occur, there are no major transformations of type. Observed change will gradually slow and cease as created potential is used up. There should be no intermediate forms between types.

As Gould and Eldredge point out, neither prediction is well sup-ported by the fossil evidence. Rather, the evidence shows long periods of morphological stasis interrupted by occasional periods of rapid morphological change and species formation. This is the pattern that they termed punctuated equilibrium. Thus the evidence reveals both stasis and rapid change—a mystery.

Let us illustrate the pattern with an example. B. Michaux, a scientist studying marine gastropods (snails) in New Zealand, has been working with fossil assemblages of the gastropod genus *Amalda* off the New Zealand coast.[6] He reports that such snail species typically remained unchanged for around fifteen million years. (By "unchanged" he means that the shells of the living animals cannot be distinguished from those in a box of their fifteen-million-year-old fossil ancestors. The two groups of shells have both the same average shell measurements and the same amount of variability.) Furthermore, during an episode of environmental change, the fossil record indicates that the snails started to change their shape, only to return to their original forms as soon as "normal" (previous) conditions were restored.

Such long periods of morphological stasis seem typical of most species. Even in the most changeable groups (mammals and insects, for instance) species typically show no detectable change for at least two or three million years. More stable groups (such as the clams) typically are unchanged for ten to fifteen million years. In fact, some organisms (lungfish and horseshoe crabs, for example) remain essentially unchanged for hundreds of millions of years. Such stasis supports the predictions of Special Creation. It does not match the predictions made on behalf of Neo-Darwinism. On the other hand, the fossil data for intermediate organisms and periods of rapid change do not match the predictions of Special Creation.[7]

So there is evidence that species stasis occurs, but there is also evidence for rapid change. Species may seem static, but they do not lack hidden genetic variability. Laboratory and field studies have shown that natural populations have vast stores of diversity. "Artificial" selection can push character traits ten standard deviations in many directions—for instance, in the oil content of soybeans or the number of bristles on a fly's abdomen.[8] Furthermore, intense selection acting on one set of traits sometimes "releases" hidden genetic variation in other traits. For example, selection for tame behavior in foxes allows the appearance of a number of traits common in domestic dog species (lop ear, spots, tail carriage, and others).[9] So change can indeed happen suddenly.

But can it cause rapid species formation? There is compelling evidence that it can. For instance, "forensic" DNA evidence shows that scores of Cichlid fish species with very different and very specialized characteristics have arisen within the same Rift Valley lakes in East Africa from a single pioneering population in the last few (perhaps no more than ten) thousand years (since the last ice age). Even a sand bar is sometimes enough of a barrier to breeding between two Chiclid populations to allow them to form different species.[10]

Another example is the anoles (American chameleons) on the islands of the Caribbean.[11] These lizards are differentiated into several species exploiting different habitats (ground, arboreal, and others). Are the various species of each island divergent descendants of a single immigrant population, or is each descended from a different immigrant? Since the arboreal species, for instance, of different islands were very much alike, it was assumed that they were all descended from a common arboreal ancestor. But that was before DNA studies. Their DNA indicates that each island was colonized only once. All the species on each island are closely related, no matter how different they look, and no matter how much they resemble species on the other islands. One colonizing species split and diversified into many species, filling the empty niches. So, then, new species sometimes do form quickly, and as they do, morphology can change rapidly. The speed of morphological change in the Cichlid fish or the Caribbean anoles is too fast to show up in the fossil record. A rapid species forming event such as this in the past would have resulted in the sudden appearance of a new species in the fossil strata.

But millions of years of biological stasis is about as unexpected as technological stasis in an industrial research and development lab. People would get fired. The firm would go bankrupt. Species would go extinct. Could the environment cause "species stasis"? But even in very stable environments, some variant should always be marginally better at getting resources, creating change. So what sort of biological knot could tie down a species? Does God prune the innovators? Are they already at the peak of perfection? Do they have genetic barriers (or "genetic cops") built into their genomes?

Old Genes, New Forms
What sort of evolutionary mechanisms will produce a capacity for rapid change, and also produce the pattern of stasis?[12] Of course, if a population has a genome already loaded with useful variation, natural selection could easily and quickly handle the process of dividing it into various descendant populations. It could sort it, but it could not produce it. And naturally, mutation can produce variation, but since it has no directionality, such a "rich" genome would have to occur by a fluke. And a "fluke" of that magnitude seems essentially impossible, given the size of the genome.

God could do it, but for various reasons both materialists and theists seem reluctant to consider this option. Perhaps there is some prejudice against hiring a spiritual being for such creative tasks. So what are the alternatives? An area of genetic probability space that happens to have new morphologies built into it seems unacceptable. Such complex, preprogrammed genetic potentials certainly are not what one would predict from a Blind Watchmaker. Hiding the specifications for new species "inside" each other like Russian nesting dolls would make selection very efficient, but it would create a real mystery. Real magic is not pulling a rabbit out of a hat, but rather inserting a rabbit into a hat, unseen. Selection can pull the rabbit out, but can mutation put it in?

We need to consider again the internal homeostatic nature of biological systems. Homeostasis requires programmed set points, negative feedback loops, and correction mechanisms (similar to a furnace being controlled by a thermostat). Such biological control programs are fundamentally genetic.

The more we learn about genes, the clearer it becomes that the differences between organisms are not a matter of which genes they have. (Human beings have 98 percent of the same genes as a chimpanzee.) Rather, the difference is a matter of which genes they turn on, and when they do it.

Genes are like computer chips. Different computers have different programs, even if they have the same chips. Different organisms have different programs, not different genes. A genetic computer is programmed by changing the pattern of information flow between the genes (chips). Genetic programming uses control elements termed "cis-regulatory sequences" (also called "enhancers"). These elements act as switches to turn specific genes on and off. This enables specific genetic elements to be changed as modules and to be fine-tuned against each other, maintaining programmed functional "set points." Such a system has intrinsic stability, being buffered against change.[13]

But not all would agree that there is a real "program." The alternative is that such "set points" are just accidental, and are temporary points where antagonistic genetic forces balance each other. Such a genetic balance is termed "pleiotrophy." But in pleiotrophy, one should still see slow change rather than stasis, because variants of the genes ("alleles") would still continuously replace one another.

Complete stasis suggests that the genome is more like a real "program," able through feedback loops to maintain specific "normal" states. Obviously, the complete reprogramming of such a genetic computer would require a great many simultaneous mutations. But selection could still occur. Changes that would be favored would be mutations reducing the cost of homeostasis, reinforcing the control loops, and thus stabilizing lineages. A million years of morphological stasis could do a lot of genetic fine-tuning![14]

What are the implications of all these observations? Is there something wrong with evolutionary theory?[15] After all, long-term stasis is not consistent with Neo-Darwinian predictions. Does extended stasis demonstrate that the mutation/selection mechanism does not work? Have preadapted complexes been inserted into the genetic hat? Could "hidden" species-level genetic programs arise spontaneously without selection (by definition, a "hidden" program is invisible to selection)? Can they be mutated into existence, fully formed, all at once? Could one even know whether such hidden programs exist? What would be the evidence of their presence? Might the creation be designed to evolve? Does this prove Special Creation? Clearly, biological reality is more complex than many have thought.

An Example of an Adaptive Radiation

The real proof of the pudding for questions such as those raised in the preceding section rests with the pattern of events that occur when new sorts of creatures appear in the fossil record. Very rapid appearance of new morphologies would tend to support the existence of hidden genetic patterns. However, if the patterns are being simultaneously formulated and evaluated by only random processes and historical contingencies (mutation and selection), the establishment of a pattern can be expected to be slow. But how slow is slow? And how fast is fast? Consider some examples.

Diacodexis, the first of artiodactyls (the name means "split hoof") appeared suddenly with a key adaptive feature, the astragalus, a bone in the ankle that allowed the feet to swing freely for fast running. *Diacodexis* was quickly joined by a whole series of familiar and unfamiliar forms that shared this unique foot structure (for example, camels, pigs, and deer)—fifteen new families by the end of the Eocene.[16]

These new artiodactyl families showed numerous parallel trends. For instance, the advanced "pecoran" foot (in which the side toes are lost) as seen in today's cows is also seen earlier in the giant pig-like creatures called Entelodonts, and it also appears early on in the camel family (although later camels lost the pecoran foot). Likewise, high-crowned hyposdont molars adapted for grazing grass (we see them in today's cows) appeared in several other line-ages—for instance, the rather pig-like oreodonts.

Such rapidly forming divergences usually are labeled "adaptive radiations" (or simply radiations, as defined above), but that may be a misnomer. Rather than being formed by selection due to the environment (adaption), the pattern looks more like the rapid unpacking of the diverse genetic potential of an existing package. But a genetic explanation requires a genetic source. The usual sug-gestion—that a key adaptation opens up a new set of possibili-ties—does not easily explain the fossil pattern. It does not provide enough morphological guidance. For example, the pecoran foot (with only two toes—the lateral toes having been lost) appears suddenly in several distantly related lineages with very different environmental niches. This suggests that they all had the foot blue-print tucked away rather than deriving it by independent adapta-tion. So does that imply that the pecoran foot was hidden in "grandpa" *Diacodexis*?

The unrolling of the potentials of the artiodactyls apparently had a major interruption: whales. A series of genetic comparisons of artiodactyl families and whales has been done. The sequences of their DNAs have been statistically analyzed. And a cladistic analysis has been made of unique bits of inserted DNA termed "SINEs" (interspersed elements). DNA results indicate that hip-popotami are more closely related to whales than they are to any of the other artiodactyls. Furthermore, cattle and deer seem to be genetically closer to hippopotami and whales than they are to pigs or camels.[17]

How fast did whales develop? There was a time when we thought we knew. Fossils seemed to indicate that whales developed during the Eocene (at the same time as *Diacodexis*) from a group of condylarths, primitive ungulates (hoofed mammals) called Mesonychids. There are a series of early Eocene forms: terrestrial *Pachicetus*, otter-like *Ambulocetus*, seal-like *Rhodocetus*, and fully aquatic dolphin-like *Dorodon* by the late Eocene. This is a beautiful example of a series of intermediate fossils. They are clearly a single, long-snouted lineage showing transformation in stages.[18]

But the DNA told a very different story. Let us begin with the DNA evidence for the artiodactyls. DNA sequence comparisons support the fossil evidence that pigs and peccaries split long ago, in the Oligocene. They match the fossil evidence that camels went their own way even earlier, back in the Eocene. But whale DNA indicates that whales separated from hippopotami millions of years later, in the Miocene. (The molecular distance between hippopotami and whales is actually slightly less than the distance between the Cervidae [deer] and Bovidae [cattle], two split-toed groups that also became separated in the Miocene.) Does this mean that the archaic whales that lived long before (during the Eocene) were unrelated to modern whales? There is not much evidence of whales during the period in between (the Oligocene). Not only is the Miocene later than the archaic fossils, but also it is when the adaptive radiation of modern whales occurred and it is the era of the earliest hippopotamus fossils. But then, in 2001, more complete fossils were found of *Pachicetus* and *Rhodocetus*. Both animals had the unique astragalus of the artiodactyls. They apparently were related to *Diacodexis*, not to the Mesonychids.[19]

What is the significance of all this detail? The fossil evidence strongly suggests that certain fully terrestrial ungulates were changed to fully aquatic whales over the span of eight million years, whereas the typical mammal species exists five million years with no discernable change. How could such a morphological "great leap forward" happen so quickly? Did whale blueprints exist "in principle" in artiodactyl genomes? Or was the new morphological program created practically overnight? God could do it either way. But how much faith must one have in undirected material agency to conclude that this could happen "on its own," with God absent? I wonder if anyone but a "true believer" in materialism can accept *undirected* material causation as the source.

Missing Links and Adaptive Radiations

So, then, punctuated equilibrium does mean both stasis and rapid change, and there is evidence for both. But does this evidence prove the existence of a designer? Can we prove or disprove design? Can we demonstrate nature's autonomy? Are one's conclusions a matter of summing up the evidence, or are they simply a matter of faith?

There are many other physical examples, but there is little point in going over them. In a war of explanations, each side presents evidence, and the other side reinterprets it. Unless we can agree on our paradigms, our way of seeing reality, we get nowhere. But we do need to consider the implications of the accepted mechanisms called "evolution" with care, for advocates of the Blind Watchmaker hypothesis intend to render God an unnecessary hypothesis.

The key issues are one's assumptions about God. How could evolution-as-relationship exclude God? Does not God govern history? Would God govern human history and leave nature to itself? A picture of change over time in a lineage is in no sense a denial of God's creative power. Rather, it is merely a description of the fossils that scientists have found. It can be proposed as a description of God's power at work. In no sense is it an affirmation of the supreme efficacy of autonomous evolutionary mechanisms.

CHAPTER 11

All Life from a Single Source?

D OES THE EVIDENCE FOR INTERMEDIATE FORMS MEAN THAT WE must accept the idea that all the diverse forms of life— aardvarks and oak tree, monkeys and mushrooms—are related to one another? Certainly, the amoebae-to-humans transformation is a somewhat more difficult pill to swallow than that of brown bears becoming polar bears, or even dinosaurs becoming birds. However, to properly evaluate the evidence for this idea, we again need to focus on what it means to say that biological things are alike or are different.

At first the answer seems obvious. Things are alike when they have the same sort of teeth, or the same number of toes, or skulls of the same shape. The problem comes when we add these questions: *How* did these things come to be alike? *Why* do these specific creatures have certain teeth, toes, and skulls?

As we have seen, each creature has an internal blueprint, a pattern for its characteristics, a body plan written upon its DNA strands. Therefore, creatures that are alike should have similar blueprints. But how did they get their blueprints?

Let us begin by considering a few more ideas related to genetics. Remember that DNA is a molecule upon which messages (blueprints) can be written. Genetic messages are written in a "language" that, as with human language, has the linguistic equivalent of letters, words, sentences, paragraphs, and eventually complete stories. The significant observation is that *all* living creatures on earth, from amoebae to humans, seem to use the same "DNAish" language. So although the differences in their bodies display the differences in their stories (DNA blueprints), all the stories are written in the same language. Based on this reality, universal common ancestry seems plausible.

Certainly a writer (such as God) might independently write many different individual stories. But a writer can also edit and rewrite. How can we tell whether similar stories of life are revisions or new compositions? We must examine them in the same way we would evaluate human literary works: we look for evidence of a history of editing (or the lack thereof).

Of course, a writer might deliberately use the same major themes in different books. Therefore, identifying common major themes is not necessarily the definitive proof of a history of editing. However, a skilled writer is very unlikely to reuse typos, awkward phrases, or long stretches of identical prose. Such elements point to a history of recycled manuscripts. Also, remember that an edited DNA manuscript may produce a change in the body described. Thus we can look for signs of a common history both in the bodies of living things and in their DNA sequences.

With millions of existing species, one could compare creatures forever. The goal here is not to provide a complete history of life, but rather to consider the evidence for a few cases of common ancestry. Thus I will confine the discussion to two examples. For genes, we will examine the previously mentioned pseudogenes and Hox genes. For bodies, we will consider the native mammals of Australia.

Evidence for Common Genes

The whole matter of biological relationship has been clarified in recent decades as scientists have learned to analyze the DNA text directly. Analyzing a genetic manuscript is in some ways similar to analyzing a literary manuscript. For example, one can distinguish various drafts of a manuscript based on when changes—edits— were made, changes that show up in all the later drafts.

With the DNA "manuscripts" of living things, the retained changes are in most cases very minor, insignificant. They do not change the meaning of the text—that is, the way the organism looks and functions—just as a literary manuscript would be essentially the same if a colon were substituted for a dash. For instance, with DNA, since there are alternate or synonymous codes for some amino acids, they may be switched "silently," with no change in the genetic blueprint. But the researcher can still sequence the DNA and determine the change. One place where such "meaningless" DNA change accumulates rapidly is "pseudogenes," extra deactivated copies of functional genes. No change in a pseudogene will affect the organism, thus such mutational changes rapidly accumulate over time. By comparing how many differences have accumulated in the sequences of the pseudogenes of two organisms, the researcher can estimate how much time has passed since a common ancestor of the organisms shared the common ancestor of the pseudogene.

Another example of a genetic "fossil" is produced by certain common viruses called "retroviruses" (HIV is the most familiar retrovirus). These viruses are termed "retro" because they work backward: they inset a copy of themselves into the DNA of the infected organism. The copy could go anywhere, into billions of different possible spots on the DNA, and frequently they will be deactivated and passed on as a "junk gene" to all the descendents of the original host organism. Any two organisms with such an "inclusion body" (a deactivated retrovirus) at exactly the same spot in their DNA must have inherited it from the same infected common ancestor.

Different species frequently have the same retrovirus "remnants" at exactly the same spots in their DNA. That would happen only if all the individuals of both species were descended from the same infected ancestor. This is one of the evidences for the Polar bear relationships discussed in chapter 9. An intelligent designer might choose to leave a harmless junk retrovirus sequence in a genome that was being edited, but it is hard to imagine why the designer would deliberately put the same piece of junk in the same spot in two separately created kinds. The explanation that makes the most sense is that the insertion took place in a common ancestor.

Thousands of molecular traces of history, such as synonymous coding changes, have been analyzed, producing elaborate similarity "gene trees," which can be interpreted as genealogies. The process is so regular it has been termed the "genetic clock" effect. The technique has been applied to everything from the HIV viruses of a single patient, to the relationships of whales, to the entire animal kingdom, to the whole world of life. How the analysis is done is very complicated. The point here is that minor DNA "edits," though they do not affect an organism's appearance or function, enable scientists to determine an organism's lineage and to determine, among other things, which living things had common ancestors and when.[1]

An "editing" history in the creation of the blueprints indicates a common ancestor, and there are thousands of examples of such realities. The logic is that retained "typos" do not make sense if each lineage was produced as a unique, "creative" effort. This is the case whether the ultimate "creative force" is intelligent (God) or undirected (chance and necessity). History is history.

Of course, it is not only DNA errors that link living things. Besides the common genetic language, there are other amazing parallels in the genetic story that link creatures that appear to be very different. One of the most surprising is the previously discussed set of genes known as the Hox group. These genes act as sequential spacers in a developing embryo, communicating to cells their location and their specialized task. For example, Hox genes tell a cell whether it is located in the head or the tail.

As we noted previously, mice and flies have parallel sets of matching Hox genes. In fact, they are so much alike that a "transplanted" mouse gene will turn on gene programs in a fruit fly. For instance, the mouse gene Hox 1.5 will cause a fruit fly embryo to grow legs, instead of antenna, on its head, which is the same effect as the mutant fly Hox gene Antp. In the mouse, this gene instead turns on the genes in the first segment of the hindbrain. They call for different programs in the two embryos, but their tasks are at the front end of their embryos. Hox-like genes have also been found in yeast, fungi, flowers, and elsewhere.[2]

The same sort of parallel genes have been found in many other locations in various genetic stories. The Pax 6 gene, for instance, is involved in eye formation in a wide variety of animals. The genetic text, therefore, both in copying errors and in textual "conventions," is compatible with a common source for all life.

But would an intelligent designer use these patterns? Who knows for sure? Whether or not the reader can follow the somewhat technical details here, the main point to grasp is the existence of a common genetic language despite the differences in body structure among different organisms. The evidence cited so far does not prove common ancestry, but it does establish that common ancestry, from a scientific point of view, makes a lot of sense.

Of Wombats and Woodchucks

So beyond a common genetic language, various details of the genetic stories also indicate an editing process, as opposed to a variety of new "creations". Now let us examine the products of the stories: patterns of body formation.

We do so by focusing on Australia. Australia is an island continent. Apparently, it has been isolated for some sixty million years, since the time of the dinosaurs. Almost all of Australia's mammals are marsupials, similar to the American opossum. The term "marsupial" refers to their abdominal brood pouch. Unlike placental mammals (the mammals of the rest of the world), marsupials produce eggs that hatch in the oviduct. No placenta is formed. The newly hatched young, almost embryonic, must climb from the birth canal to the pouch. They will remain there developing, locked to a nipple, for the next several weeks or months.

Marsupials share various additional characteristics that distinguish them from placental mammals. For instance, they do not replace the milk teeth, but rather add more teeth behind them. So, then, if all forms of life are related, the bodies of marsupials seem to be off on a side branch that separated long ago (about one hundred million years) from a common branch leading to all the placental mammals, including aardvarks, whales, bats, tigers, and humans.

One therefore would expect the Australian mammals to have forms very different from the various placental types. However, this is not the case. There are marsupial look-alikes for badgers, pine martins, weasels, wolves, mice, flying and nonflying squirrels, lemurs, moles, mice, jerboas, rats, rabbits, woodchucks, anteaters, and more. There are even the fossil remains (and occasional reports) of a lion-like marsupial "cat." The familiar kangaroo is the functional equivalent of a deer, despite its mode of locomotion, and there is a placental equivalent with kangaroo-like locomotion: the springhaas of Africa. Marsupial lineages share a host of specific adaptive structures with their placental equivalents: sticky ant-catching tongues, carnassials (meat-slicing cheek teeth), gliding membranes, digging claws, even saber teeth (in an extinct South American form), to name a few.

What can we deduce from this "alternative zoo"? First, note that quite a few different major mammalian themes were worked out in both placental and marsupial forms. Yet despite a number of rather complex parallels in adaptive structures, species remain distinguishable as marsupials or placentals. Diverse themes are being played out in each area against an underlying local structural unity.

Second, high marsupial diversity is seen only in Australia, where there are no living or fossil native placental mammals. Most carnivores apparently went extinct when the native Australians introduced a placental canine, the dingo, to the continent. The dingo became the Australian continent's lone large carnivore, apparently out-competing its rivals. Although marsupial fossils have been found around the world, no other continent or island has any living species of marsupials, except for a few species of opossums. The fossil record of South America did display a radiation of marsupial carnivores, which went extinct fifteen million years ago at the time placental carnivores arrived from the north (after the Isthmus of Panama reconnected North and South America).[3]

What does all this suggest? The common features displayed by these two geographically isolated groups of species are very difficult to explain unless each group of species is descended from its own common ancestor. Could God have directly created Australia's creatures with these peculiarities? Of course. If so, for some reason God used a few odd and unique structural details in a wide diversity of creatures found on a particular land mass. It looks like common ancestry—diverse species molded from the same living clay.

On the other hand, the many detailed parallels between specific marsupial species and specific placental species are very difficult to attribute simply to two distinct common ancestors. The pattern looks more like a designer carving common themes out of two different batches of similar clay.

Would such a designer have to be God? If so, we have just proved God's existence. But there are three other candidates that have applied for the job. Following are their qualifications.

The first designer is Darwin's candidate. The environment "selects" variants to match the various functional niches in the ecosystem. It looks good for explaining basic functional items, such as why dolphins and fish are streamlined torpedoes, but we are talking about some really detailed structural parallels here. It is not clear that these were the only ways to fill those niches.

The second possibility is that the common information found in the genome of both of the ancestral populations was detailed and constraining enough to restrict access to only a few blueprints, the body forms we see. Today's species thus were hidden in their ancestors' baggage. The question is whether they could they fit into the bags. Does not selection provide a "flight limit" on hidden complexity?

According to the third candidate, diverse packets of new information, new blueprints, were at some point inserted into the two lineages by some outside common source. Perhaps these new blueprints rained down after being formed on a comet, or were seeded into the atmosphere from a spaceship sent by a distant civilization. Fred Hoyle and Francis Crick, the co-discoverer of DNA, actually proposed these as nondivine scenarios.[4]

Some would maintain that if God did indeed intervene, the only logical method of the three interventions would be the third: inserting diverse packets of new information, new blueprints, into the two lineages. But why would it be assumed that God could not be equally involved with the first two possibilities? Is not God the sovereign who governs the selecting environment? Would God not be the one responsible for packing the genetic bags of the first Australian mammalian immigrants? And when God does insert material into the genome, is it not divine choice as to whether it be accomplished all at once or in smaller increments? Might not mutations be a sort of divine form of genetic engineering?

In any case, it seems certain that some unusual "creative force" produced the various Australian lineages as edited versions of an original marsupial text. Believing that the creative force was God, or that it was unguided nature, is a choice of faith, not science. For my part, I am convinced that, based on the probabilities, believing that unguided nature was responsible is a blind leap. But some materialists have great faith.

Resolution: Genetic Similarities

We have focused on only a few cases. There are many, many others that I and other biological scientists have studied. And as we continue studying and learning from nature, what if the pieces of this immense jigsaw puzzle continue falling into place in ways that connect bacteria to fungi, oak trees to mosquitoes, all of life into one genealogical web? The likely conclusion would be that indeed all living organisms have a common ancestor.

But which of the four causes would this history of life illuminate? Common ancestry is about *material* cause, the stuff of which things are made. It does not say what forces (*efficient* cause) might cause descent with modification. It does not answer questions about design, about whether a common plan is being worked out (*formal* cause). And certainly it does not answer questions about whether there is purpose in their creation (*final* cause). It could tell us neither the purpose of any species nor that it had no purpose.

Certainly it could not directly show how God governed or did not govern, except for the wonder generated by very improbable events. But there is no logical reason to conclude that the descent of existing life from a common ancestor would be possible in an autonomous universe. The reasons to believe that God was necessary for what we see would be just as strong. One final note: Since God can work via process as easily as not, evidence of process does not exclude God. Likewise, evidence of design does not exclude historical processes acting to produce the design. We have to trust God and go with the records left behind for us in God's works.

CHAPTER 12

Another Mystery beyond Explanation

THE SUDDEN APPEARANCE OF ANIMALS, COMPLETE WITH ADAPtive structures (for example, eyes, hearts, legs), is the sort of thing that intelligent-design theorists look for, hoping to find evidence of irreducible complexity, They contend that these creatures had to have been directly created because they appeared, seemingly out of nowhere, complex, complete, and fully functioning. Other than the origin of life itself, there is probably no bigger mystery in the history of life than the sudden appearance of animals during a period known as the Cambrian.

Before about 535 million years ago, there is almost no evidence of animal life. From the first appearance of animals, one would expect natural processes to require at least hundreds of millions, perhaps even billions, of years for a variety of life forms to develop. That is not what happened. As of about 535 million years ago, all of a sudden we see representatives of all the animal phyla and most of the classes of those phyla ("phyla" is the taxonomic term for the major groups of organisms; classes are parts of phyla). Can Neo-Darwinian mechanisms alone explain this fascinating mystery? Let us take a closer look.

Examining the Science

For billions of years prior to the Cambrian era, the earth's seas apparently were populated only by single-celled creatures and algae. The first sign of change is found in strata about 600,000 years old from the Edicarian Hills of Australia. This group of fossils (known as the Edicarian assemblage) consists of odd, flat-bodied creatures with no clear relationships to any modern creature, with the possible exception of jellyfish. In addition to these creatures, there are worm trails in the ancient mud made by some unknown, worm-like creature.

Then in the blink of a geological eye, all the known animal phyla appeared and diversified—a sort of biological Big Bang. There are almost no fossil hints about their ancestors. Some small, shelly fossils had been around for about fifteen million years (the Nemakit-Daldynian period of the Cambrian). But in a span of five million years (the Tommotian and Atdabanian periods of the Cambrian), all the known phyla (over thirty) appear. Five million years might seem like a long time, but as we have seen, it is actually about the length of time that the typical mammal species remains unchanged.

Furthermore, there are fossil beds that provide excellent preservation of creatures that lived before, that lived during, and that lived after the Cambrian transition. The earlier Edicarian beds do not contain any examples of the crown-group phyla (the thirty or so major groups of animals alive today). The Edicarian fauna apparently vanished before the Cambrian types appeared, but when they did, these new types of animals appeared very rapidly.[1]

What happened? Is there any physical evidence to provide clues? Only a little. In Greenland a fossil assemblage known as Sirius Passet has preserved almost the first moments of the new animal life. Two examples are of particular interest. First is the halkieriids, slug-like creatures covered with a "chain mail" of scales called sclerites, topped off with a clam-like shell on each end. Their "body plan" seems poised to become mollusk, polychaete annelid, or brachiopod; thus they could be ancestor to three of the major invertebrate phyla. Second is *Kergmachela,* a free-swimming creature with lobopod-like (unjointed) appendages on the trunk, and arthropod-like (jointed) sensory appendages on each end. In both cases, the creatures had characteristics that place them outside, but related to, more than one modern phylum. If they do represent early stem groups that gave rise to the phyla, the source of their characteristics remains unexplained. The period of their existence was very short. Modern phyla replaced them in very short order.

So the Cambrian represents something far more dramatic and far more complex than a simple "cow-to-whale" transition. At least in that case, complex body plans were available to be modified. No such metazoan body plans (genetic descriptions) were in place for the first animals. And one can neither modify nor select what is not there.

Possible Explanations

So, are there any strictly natural explanations that make sense? To some extent there are, but to accept them as completely adequate requires a great deal of faith. As paleontologist Robert Carroll puts it, "This explosive evolution of phyla with diverse body plans is certainly not explicable by extrapolation from the processes and rates of evolution observed in modern species, but requires a succession of unique events. The development of complex body plans, with many distinct cell types and anatomical structures, required a new system of genetic control that is not present in unicellular organisms."[2]

Why does Carroll say that the standard theories will not explain the Cambrian Explosion? Because to construct a metazoan (an animal with many cells) requires not only the blueprints for different sorts of specialized cells, but also blueprints that determine how the different types of cells are to work together on tasks to make the whole organism viable. To use an analogy from the computer world, new software is not enough. You also would need a disk operating system that allows the software to function. The question for science is how elaborate genetic blueprints could have emerged so quickly.

Consider the following analogy. A family lives in a certain valley and makes pottery. The parents teach their children to mold and fire pots, and the family in turn trades its pots with people from other valleys to meet their needs. Then the industrial revolution arrives in the family's valley. The family now must construct large buildings instead of pots. The children need to specialize. They must learn to work in glass, wood, steel, fabrics, plastics, and, for some of them, ceramics (for drain pipes and roof tiles), as did their ancestors. But they must be taught all these new skills by their parents, the potters. Obviously, their parents will need to be several times smarter. And furthermore, everyone's specialty has to be subcontracted to the building contractor, who is operating from an overall blueprint. And the parents have to draw up the blueprint and be the contractor as well. Unless the parents are given a crash course in all the relevant technologies, it is not going to happen.

Consider the implications of how fast morphology changed in the Cambrian. According to Neo-Darwinist theory, the speed of selection reveals the speed at which genes are being discarded from the gene pool. This in turn is determined by the amount of genetic diversity in the gene pool (a principle known as Fisher's Fundamental Theorem of Evolution). Imagine genetic diversity as the amount of gasoline in a car's tank. Since selection burns up diversity, one would expect that high-speed evolutionary change would quickly run out of fuel, slow down, and stop. But in the Cambrian it did not. Clearly, there was something very unusual about those initial gene pools.

As we noted previously, gene pools are refilled with the "diversity fuel" resulting from (presumably random) mutation. The diversity used as fuel for the change would have to have been present in that primordial Cambrian gene pool. But how could a single gene pool contain the instructions for all thirty phyla? Neo-Darwinism typically assumes that evolutionary change is driven slowly by selection choosing from minor changes in the gene pool. This provides time for small-effect mutants to refuel the gene pool. Certainly, one would expect it to require a far more extended period for evolutionary development. Again, such random, slow-paced refueling cannot explain the explosive appearance of over thirty different body plans in five million years if they had only one common ancestor, not many.

DNA comparisons within the last decade have led to a new classification of all animal phyla into three major groups. Each group includes both simple phyla and complex phyla. The DNA indicates that the three groups separated about half a billion years prior to the Cambrian, or about one billion years ago. And since the complex phyla in each group have numerous common characteristics, their common ancestor must have been complex.[3]

Amazingly, the fossil record shows no sign of this secret history. No physical evidence for over thirty lineages of pre-Cambrian animal life exists—a mystery that is mind-boggling to a scientist. So where were these ancestors? Is it here that we see that intruding divine finger? Well, that is the problem of making an argument from missing evidence. As I edit this manuscript, I find that the missing probable ancestors have just been found in China (in the Doushantuo Formation)—ten specimens of a tiny coelomate, bilateral "worm" named *Vernanimalcula* that apparently lived about fifty million years before the Cambrian.[4] Furthermore, at the same time, I find that the genetic divergence has been recalculated due to the realization that the genetic clock is slower in vertebrates than in invertebrates.[5] The new calculated time of divergence is also about fifty million years before the Cambrian.

But much of the mystery remains. If the rapid change of the Cambrian was driven by new mutation, genetic probability space must have been absolutely loaded with gold. Or rather, the genome of the common ancestor was stationed at that really unique spot from which all the body plans could unroll. We know neither if such a universal genetic junction exists, nor if it does, how it could be found or reached.

Some materialists say that, improbable as this explanation is, this is what *must* have happened. But such a conclusion is determined by the premise. If one begins with the assumption that only matter is autonomous, then no other explanation is possible. From the vantage point of science, however, the Cambrian Explosion does indeed show that as a mechanism, an unguided, "Blind Watchmaker" form of Neo-Darwinian theory is inadequate. Gradually accumulating a random bunch of small-effect mutations does not explain the formation of new body plans overnight. Something of larger scope must have been involved.

But we must be careful about drawing conclusions. It may be tempting now to assume that we have spotted the finger of God, acting apart from secondary causation, intervening in nature to build new bodies at a rapid pace. But science has some theories related to Hox genes, regulatory networks, and genetic feedback that address the mystery of the Cambrian Explosion.

Hox Genes and Common Ancestors
Let us examine this idea of an ancient, common ancestor for the Cambrian forms that was too camera-shy to show up in the fossil record. Have we any evidence of this mystery beast, or are we just writing on the edges of our maps, "Here be dragons"?

Actually, the existence of this ancestor is not based totally on speculation or desperate theories. In a previous chapter we noted the exciting recent genetic discovery that all metazoan animals apparently share matched control genes, the Hox genes. In species from fruit flies to flying foxes, the same sets of Hox genes lay down the body pattern, head to tail. They control the specialization of tissues in time and space, turning on further genetic "packages" of other control genes "shared" in flies and mice. For instance, "pax 6" makes eyes, the "tinman" gene makes hearts, and "math 1" makes ears (auditory mechanoreceptors) in both flies and mice. This, of course, suggests that an unknown common ancestor possessed these genes. Is it possible, then, that the appearance of control genes such as the Hox genes in the common ancestor made possible the Cambrian Explosion?

Even this explanation for the Cambrian Explosion, however, takes something for granted. The function of Hox genes is to turn on genetic programs. They cannot write the genetic program, without which they cannot operate.

Perhaps there was an Edicarian-era "urbilateralia"—perhaps *Vernanimalcula* (we probably would call it a worm) was equipped with Hox genes ready to spin out the diversity of the Cambrian. In that "worm" the metazoan body plans already existed. They needed only to be released and expressed. Indeed, Knoll and Carroll suggest exactly that.[6] The poor urbilateralia was kept under, unable to express its glorious potential due to the dead hand of the Edicarian ecosystem. However, when a major extinction event removed the dominant Edicarians, the urbilateralia were "released" to diversify gloriously into the Cambrian. Accurate or not, it surely is a great story! The problem is the improbable amount of morphological diversity that must be in waiting in that common ancestral population.

There are suggestions for the origin of this variation, of course. These ideas are far from simple. For instance, Carroll suggests a change in embryological development. In many small invertebrates the embryos develop directly into adults with no larval form. In this "type I embryogenesis" (embryo formation) developing cells have absolutely determined fates. It produces very small feeding animals with a "basic" anatomy of gut, muscle, nerve, and epidermis.

Davidson and others argue that this ancestral developmental mode is not sufficient to generate large animals. He thinks that the key innovation that caused the Cambrian Explosion was the appearance of brand-new "set-aside" cells in the larvae that could respond to new control genes (like the Hox set) and be used to construct new structures in the adults ("type II embryogenesis").[7] But again, the genetic control programs still need to be written—something that Hox genes cannot do.

Most forms of life have very similar genes. Where, then, is the location of the variability that allows selective change? If it is not located in the genetic loci themselves, it must be located in the signaling connections between them. The proteins produced by most loci (genes) are transcription factors—signals to other genes. A very large part of the so-called "junk" sequences outside the loci are likely to be enhancer sequences—the signal receptors that recognize the transcription factors. David Stern of Cambridge University maintains that most of the diversity that determines body form and function (and fitness) resides in these interactions.[8] That is, the information is in the feedback loop. Significant mutations, those that produce the variation that is likely to be selected, occur in those cryptic locations, at the reception sites of the genetic feedback loops. But very little data has been collected concerning the nature and selection of such variation.

Such feedback loops do allow for the possibility of something called "evolutionary feedback." According to this concept, a mutational change in one loop produces an altered trait that is favored, thus increasing the likelihood that future changes would be selected that would stabilize this new variant form. Such feedback would account both for the extraordinary stability of the usual periods of stasis and for the rapidity of change when it did occur. Populations undergoing such periods of rapid change would also be fragile, thus likely to disappear. Thus change would not happen very often. In theory, the mechanism could produce the pattern previously described as punctuated equilibrium, though the Cambrian is a pretty extreme example. And it has not yet been demonstrated to be an effective mechanism.

God's Role

Strictly materialistic evolutionary mechanisms do have their problems, and they are critical ones at that. Adaptive radiations and the appearance of new forms seem to happen too fast to be accounted for through undirected mutation plus selection. True, there are some theories, but they depend on probabilities that are so hard to "go figure" that resting in their assurance is a matter of pure faith. It certainly is astonishing that some people believe that science has demonstrated a mechanism that without guidance can produce biological reality. How can one reject the possibility of guidance when one does not understand the background probabilities? The fact that we have walked through the forest in safety does not prove that we had no protection and that there was no path.

One obvious conclusion of this discussion is that arguing at the cutting edge of theoretical evolution is hard mental work. But the new mechanisms being proposed are no better than the old at excluding the hand of God or predicting the likelihood of these events. And the ancestral fossils are still quite rare. Whether or not this is the correct (or even a possible) mechanism for explaining the events of the Cambrian, complex and rational explanations have been proposed and will continue to be proposed. Thus the Cambrian cannot be used as a sort of "blanket proof" for miracle. The mystery remains, but the detectives are still on the case.

Finally, what should Christians think about these issues? That is not so hard. God is free to do it either way, and he alone knows the right answer to the mechanism question. Whichever way God did it, we can trust him. But clearly, whether we would like to see sudden miracles or gradual governance in creation, it seems quite rational to conclude that divine intelligence was behind it all. Yes, this is the response of faith, but on this point, accepting the Blind Watchmaker philosophy is also a faith response, not science.

CHAPTER 13

The Origins of Human Beings

I N PRECEDING CHAPTERS WE HAVE LOOKED AT STRONG EVIDENCE for the common descent of life and for reasonable genetic mechanisms to produce change. In other words, the evidence strongly suggests that evolution, as it is generally understood, did take place, that over the course of hundreds of millions of years all the forms of life that exist today descended from a common source.

On the other hand, we have seen that there is no credible reason to assume that evolutionary mechanisms without guidance could have produced the world of life that we see today or explain the events of the fossil record. To put it another way, for those who are viewing evolution from the vantage point of Christian faith, it is not hard to see God's guiding hand at work. But it is the kind of evidence that, while pointing toward God, cannot prove God to a devoted materialist.

Indeed, the forces proposed to act in evolution are claimed by the Scriptures as being under the power of God. In other words, the scientific evidence neither makes the materialistic philosophy termed "naturalism" logically necessary nor renders it more persuasive. Dawkins's Blind Watchmaker hypothesis is just that, a hypothesis—mere philosophy, if you will. And calling it "evolution" does not increase its scientific credentials.

Now we move to the most contentious area where evolution and creation have met: the question of the human race. Issues discussed in previous chapters attract interest largely because of their implications for the significance, the uniqueness, of human beings.

The question of human origins raises a host of new theological issues. Are we different in essence from other forms of life? Were we made in the same way? Do we have an eternal soul? How should we understand sin, redemption, incarnation? Such bedrock concepts of the Christian faith are directly affected by our understanding of our origins. And in no area concerning the beginnings of things does the Bible have so much to say.

The followers of the naturalism faith, on the other hand, are convinced that all human characteristics, like those of all other creatures, are ultimately shaped by the need to survive and leave offspring. Thus they are busy explaining sexuality, diet, language, consciousness, perception of beauty, the meaning of truth and rationality, and so forth as the historical products of mechanisms that enhanced our chances of leaving our genes. Frequently, it seems that many of the "new" answers they come up with suspiciously resemble popular behaviors forbidden by traditional moralities—for instance, a justification of sexual promiscuity as an evolutionarily selected adaptation.

It is understandable, then, that a person wanting to defend children against the lure of a life void of Christian values would argue in a strong and simple fashion for the creation of humans. Can we find Adam and Eve in the fossil record? Is the human record free and clear? Consider the following statement from a presentation for youth: "Instead of thousands, or hundreds, or scores of examples of links between beast and man, we find not one undisputed example."[1] This is easy to say, easy to understand. But is it true? Is there really absolutely no hard evidence supporting the river of books, television specials, and magazine articles about the "caveman"? And if that fossil evidence does exist, are our children safer if they are protected by ignorance? Dare I say, if they are protected by a lie?

This chapter will examine the data in three main sections. First, we will focus on the bones. Are there fossils of creatures intermediate between beasts and people? If so, what should we make of them?

Second, we will examine evidence from that past that pertains to the *imago Dei,* humans' unique status as bearers of the image of God. What does it mean to be a bearer of God's image? How is it possible to understand the life and abilities of people/creatures that left almost no evidence of how they lived?

Third, we will look at genetic blueprints. The recent explosion of genetic information has given us a great deal of new information about the nature and history of the human species. This evidence will clarify the unique biology of the human race, but it will also show our "creaturehood."

Then, finally, we will review these various strands of evidence in light of who God is and what the Scriptures say.

Missing Links

Looking for missing links entails doing forensic detective work. As in all forensics, we cannot go back and watch our ancestors. If that is the sole requirement of proof, then nothing from the past can ever be proven. Whether it was "the Cambrian Explosion," "Noah's Flood," or the birth of our own grandparents, we were not there. But I am convinced that examining evidence of past events can lead to very high degrees of certainty, whether we are examining fossil evidence or investigating a murder.

So what is the evidence? Let us begin with anatomy. Obviously, certain animals are more human-like (hominoid) than others, most notably those species that we call the great apes. If they and we descended from common ancestors, fossil bones of intermediate creatures might exist—"missing links"—a trail of evidence leading back to that common ancestor. And such bones do exist, lots of them. There are many skulls, but also some fairly complete skeletons. The question is this: What were these creatures whose bones we have found?

In general, most of these old bones fall into two groups. One group, found only in Africa at sites that have been dated to as early as six million years ago, are from extinct creatures called the Australopithecines. The best-known Australopithecine find (in Ethiopia), dubbed "Lucy," has a fairly complete skeleton. There seem to have been several Australopithecine species, perhaps as many as six or eight, signified in part by different sizes and weights. Earlier specimens are more "primitive" (ape-like). Later specimens have been divided into groups with heavier jaws (robust forms) and those with lighter jaws (gracile forms).[2]

Australopithecines could best be considered apes that had been modified for upright walking. There are a few significant differences between these fossil species and the modern chimpanzees. Australopithecines had slightly larger brains, shorter canine teeth, and a pelvis more suited for upright walking. On a continuum between chimp and human, they would be about 30 percent closer to the chimp. The ape most like these skeletons today is probably the pigmy chimp, or bonobo. Certainly, these forms are the intermediate ones we would expect to see if an ape lineage were to develop in a human direction. But they are in no sense human.

The second group of old bones is more human-like, and it is placed with us in genus *Homo*. The earliest of these bones have been dated to at least two million years ago. They have been found at very early dates (over 1.5 million years ago) in Africa, Asia (Indonesia), and Europe (Russia). The most complete skeleton is that of a young male found at Nariokotome in Kenya.[3]

From the neck down, these creatures had a body pretty much like ours. Their bones were thicker (they were much stronger), and their pelvises were slightly narrower (smaller heads on the babies).

From the neck up, however, their skulls were different. The brain was smaller, the face was bigger and projecting, and the cranium was long and low.

The bones of the later (after 300,000 B.C.) archaic hominids were heavier, especially those in Europe (the Neanderthals), and their brains were bigger, some as large as those of modern people. However, the basic style of skull and pelvis remained archaic. In general, compared to modern humans, these hominids were more robust; had a relatively flat basi-cranium and thicker skull bones; showed smaller, lower, and more elongated cranial vaults with more buttressing and torus formation; had larger facial skeletons with larger teeth; and lacked the mental (chin) eminence.[4] Such populations persisted until around thirty thousand years ago in Asia and Europe.

But around 160,000 years ago, people with skulls, faces, and pelvises like ours showed up, fairly abruptly, especially across North Africa and the Middle East. From the first, they had the high rounded skulls, retracted faces, pointed chins, and light bones of modern populations, although initial populations were rather more sturdy than we are.

The data look like this: Modern people lived in the Qafzeh and Es Skhul Caves on Mount Carmel (Israel). But when? These populations have been dated by thermoluminescence (TL) and electron spin resonance (ESR) measurements, and by the associated small mammal fossils, to around 90 to 110,000 years ago.[5] (Both techniques depend on the capture of photons from cosmic radiation— TL in tooth enamel, ESR in unheated flint. Thus, the amount of captured light is proportional to the time since the tooth was grown, or since the flint knife was thrown into the campfire.) Surprisingly, TL dates the Neanderthal remains at the adjacent Kebara Cave to sixty thousand years ago. Thus, modern people preceded the archaic Neanderthals in this area.

Anatomically modern human remains of even earlier populations have also been found in the Middle Awash area and at Omo-Kibish in Ethiopia, and also at Border Cave and the Klasies River mouth in South Africa.[6] The earliest modern sites in the Far East are in Australia at around sixty thousand years ago. The earliest European modern peoples (named "Cro-Magnon" after the valley in France where they were first discovered) seem to have arrived around forty thousand year ago. Within another ten thousand years, the European Neanderthals had died out.[7] Their 200,000-year-long run was over.

In that sense, then, there are indeed fossils that link modern humans to apes. They really did live—creatures that were physically "in the middle" between apes and us. But what were they? What does it mean? It is not easy to say. We probably could call the various Australopithecines species apes. But what should we call the archaic *Homo* groups? Were they animals, humans, or something intermediate? Is it possible to be intermediate? To evaluate that, we need to consider what it means to be human.

The Essence of Humanity: The Scriptural Model

The Bible holds that the uniqueness of all creatures lies in God's eternal decrees. What, then, is the decree for human beings? Certainly, God's decrees are reflected in his creatures, but the Scriptures, as verbal communication, speak more directly to the essence of things. Here are a few of the relevant comments from Genesis:

> So God created man in his own image, in the image of God created he him; male and female he created them. God blessed them and said to them, "Be fruitful and increase in number; fill the earth and subdue it. Rule over ... every living creature.... I give you every seed-bearing plant ... for food." (Genesis 1:27-30)

> The LORD God formed the man from the dust of the ground and breathed into his nostrils the breath of life, and the man became a living being.... The LORD God took the man and

put him in the Garden of Eden to work it and to take care of it. (Genesis 2:7,15)

To Adam he [God] said, "Cursed is the ground because of you; through painful toil you will eat of it ... until you return to the ground, since from it you were taken; for dust you are and to dust you will return." (Gen. 3:17,19b)

So, then, the scriptural description states that, in common with (other) animals, humans are made of dust (the same material), are given the same food (green plants), and are given the same command (to increase and fill the earth). Yet humans are uniquely intended to be God's viceroy, ruling the animals and the earth (fleshed out in Genesis 2 as instructions to run God's garden, possibly to extend it over the earth). Thus human beings are made "in the image of God," a phrase echoed throughout the Scriptures. Man and woman's unique commission as secondary governors, and their imaging of the Creator, must therefore be a picture of God's unique decree for the human species, translated into words for human understanding. But what does it mean to be in God's image?

The *imago Dei* has been discussed and debated extensively throughout church history, and clearly this book can only introduce the subject. However, there are a number of major themes that continue to surface in the discussion of the image of God. We will consider each of them as a facet of the complete scriptural description of God's intent for humankind.

The first facet is reason, the concept that as a "rational soul" a human being mirrors the thought of God, that he or she can understand God and the world that he has made. Thus men and women can communicate with, companion with, and worship their maker. This view was central to Thomistic theology (influenced by the Greek concept of eternal reason). In contemporary times, theologians such as Carl F. H. Henry and Gordon Clark have subscribed to this theology. As Clark states, "The image must be reason or intellect. Christ is the image of God because he is God's Logos or Wisdom. This Logos enlightens every man that comes into the world. Man must be rational to have fellowship with God."[8]

The second facet of the image of God is righteousness. This was the understanding of the Reformers John Calvin and Martin Luther. Human beings are to mirror God's holy character in thought and in life. But unlike the rest of creation, we can choose to obey or to disobey, since we are fully conscious of our own selfhood. Humanity is fallen; the image is defaced but not completely destroyed. The fallen person still knows righteousness and is capable of rational thought. "For although they knew God, they neither glorified him as God nor gave thanks to him, but their thinking [reason, understanding] became futile and their foolish hearts [perception, will] were darkened" (Romans 1:21).

The third facet of the image of God is relationship. Humans are to mirror God (who exists as three persons in one being) in forming relationships—with God, with spouse, with friend and neighbor, with all creation. "So God created man in his own image, in the image of God he created him; male and female he created them" (Genesis 1:27). We reason or rationalize, show righteousness or selfishness, in community. The image of God is not fully expressed in a solitary life. "The preservation of humanness has often been interpreted as the preservation of understanding and will, but actually it manifests itself in a much deeper and more important way in the various sorts of relations between man and fellow man."[9]

The fourth facet is that people image God by the office that they were given at their inception. That office is dominion over the earth, and our intellectual and physical abilities can be viewed as equipping us for that office. "You made him [man] a little lower than God and crowned him with glory and honor. You made him ruler over the works of your hands; you put everything under his feet" (Psalm 8:5-6, NIV). Thus the human species mirrors the kingly activity of God in obedience to its creational commission: to govern under God and to further realize God's purposes in the creation. Human cultural activity and development were to reflect the activity, intentions, and character of God, increasing the goodness of the creation, tending and extending the garden of God (and thus to subdue the earth).[10]

If the human race was created to serve the creation by ruling it, with the fall of humankind that dominion becomes a corrupting influence, ruining all it touches. "For the creation was subjected to frustration, not by its own choice, but by the will of the one who subjected it, in hope that the creation itself will be liberated from its bondage to decay and brought into the glorious freedom of the children of God" (Romans 8:20-21). And all this is still part of the concept of who we humans are.

In each case, humanity is *commanded* to image God. Imaging God is a calling, not a static likeness carved into our flesh. Although theologians have never thought of the image as a material likeness, the material nature of human beings is not irrelevant. God does not have a material body, but since humankind is a unity that includes the material, the image of God touches on our material aspect as well. Our physical characteristics (having hands, for instance) support the task of imaging God, even when they are shared with other species.

All these ways of defining humanity hold true regardless of how our physical beings were formed. What defines the human race is something other than the physical. Theologically speaking, God ordained the human race from eternity. God in eternity determined the nature of humanity. It is reasonable to believe that this creation ordinance affected all of creation, not just human beings. After all, the rest of creation was part of the causal pathway leading to the creation of those who bear God's image. Thus we should not be surprised or insulted to find traces of human characteristics in other creatures. This does not imply that these other creatures bear the image of God, even in part. They have other callings to obey.

Some associate image too readily with visual likeness. But do monkeys look more like God than do lions? After all, they look more like us. It seems likely that is not what the term means. Does "image" mean mental or emotional characteristics? Wolves care for injured members of the pack; elephants remember their dead; chimpanzees make tools and can be taught to use symbols on a computer. Do these animals have a bit more of the image of God than other creatures? Or can we simply expect occasional reflections of God's character in a created cosmos?

Clearly, *all* the nonphysical qualities listed above are characteristic of humanity. Since humanity is a unity, all of them together are aspects of the eternal decree by which God calls humanity into existence. The typical list of unique human qualities (abstract reason, representational art, complex linguistic structure, religious belief, accumulated knowledge, cultural flexibility and complexity) are simply realizations (albeit, often distorted) facets of the image.

But how, then, did the eternal ordinance for the human being, the creature who images God, call the human race into existence? Was it instantaneous or gradual, made from like dust or unlike dust? What evidence do we have for God's method in the creation of the human being? Adam, where are you?

Looking for Humanness in the Stones and Bones

The problem with looking for the image in fossils is that most of the ways the unique human qualities are expressed seem rather ephemeral: reason, righteousness, relationship, and rule. We need concrete evidence of human thought: art that is representative or symbolic; the manufacture of objects that took significant cognitive planning; evidence of a lifestyle that displayed human qualities, including behaviors such as religious thought and commerce. But baskets and leather goods do not survive for a million years. What we have from the far past is durable goods—stone tools, bones, and pottery—the evidences from campsites and burials. It is sparse data, but there is a pattern.

There is nothing below layers dated to around two million years ago that indicates any activities other than those typical of modern chimpanzees (which modify grass stems to produce termite "fishing poles"). They were worthy animals, even using tools, but clearly not human. At around that time, in Africa, one finds stones (Oldowan cobbles) that have three or four large chips knocked off, providing a cutting edge. Presumably, they were used to cut something. The usual assumption is that the cobbles were chipped by a short-lived species that has been termed *Homo habilis,* and perhaps they indicate a shift to more meat in the diet. There are no other signs from that level.

At levels dated from 1.5 million years up to 200,000 years ago there is copious evidence of a particular type of stone tool being chipped: a pointed, flat, bifacial oval hand axe, made in what is called the Acheulian tradition. There is also evidence of the use of fire, and possibly (in Europe) shelters of some sort. These are the years of the species called *Homo erectus.* But over this vast expanse of time the "technology" did not change at all. There is no evidence of things such as burials, art, and commerce. Thus it would be difficult to consider these creatures human beings.

Starting around 250,000 years ago, tools were made by a different technique—the Mousterian tradition—that involved more reworking. In addition to hand axes, a few other shapes were made, notably a flake struck from a prepared stone blank—a process called the Levallois technique. There is more evidence of fire, shelter, and burials, but no evidence of art or ornamentation, at least in Europe, during this period.[11]

Mousterian tool-making techniques were displaced by Aurigancian techniques around forty thousand years ago. Suddenly, bone and antler are being used as well as stone. Stone tools are prepared by the disc core technique, in which uniform blade blanks are struck from the edge of a prepared stone core and then modified into specialized tools. Tiny blades are made as replaceable components of complex tools. We see new techniques, new materials, art, music, ornamentation, elaborate burials, new patterns of resource use and trading of materials over hundreds of miles—and rapid change in all of these. Suddenly we recognize ourselves. These people were human. This sudden cultural transformation of the Aurigancian in Europe has been described as a cultural explosion. What happened? The fossils indicate that modern people entered Europe at that time. Thus, the Mousterian seems to be the culture of the later archaic hominids (like the Neanderthals), and the Aurigancian to be the culture of the modern folk like us.[12]

The problem with this neat solution is that modern skeletons have been around about three times that long, at least in Africa and the Middle East. If rapid cultural advance is tied to modern-looking people, why did it not get started sooner? The evidence is that it did. In Africa, at a number of sites, there is evidence of advanced blade and bone tool production, ostrich shell beads, ocher paint, and symbolic markings ranging back to 90,000 B.C.[13] And the oldest modern fossils, the skulls of the Awash area of Ethiopia, were neither buried nor discarded. They had been carried around and stroked for a long time after death.[14] We will never know what those who kept and used these skulls were thinking, but clearly the skulls had symbolic meaning to them. Europe's abrupt record is therefore of an invasion, not an invention. Modern bodies are (at least potentially) tied to modern behavior.

Returning to locating Adam, it seems clear there have been several species of human-like creatures besides the living apes. The pattern of their appearance seems to go in steps, to be "punctuated." Changes appear relatively rapidly in between long periods of morphological and cultural stasis. But there are several such steps, not just one. Is Adam at one of the steps? And if so, which one? And also, did Adam have ancestors on the other side of that step? This is a mystery. We need some DNA forensics to determine parentage.

Humans' Genetic Links to the Rest of Creation

Modern genetic data can be used to address several questions, including this one: What is the degree of human similarity to the living ape species? When human gene sequences are compared with those of the apes, chimpanzee and human genes have about a 98 percent match-up. In fact, however one evaluates the human genome (chromosome banding, protein sequences, pseudogenes, groups of alleles, retroviruses, Y chromosomes, microsatellite DNA, mtDNA, and so on), the results always yield the same unexpected pattern. Human and chimpanzee gene sequences are closer to each other than either set is to the genes of gorillas. Furthermore, the genes of orangutans are equally distant from all three.

A specific example is the data from a paper, co-authored by A. H. Salem and several others, that analyzes retrovirus elements called Alu elements. All humans share seven identically placed Alu inserts with all chimpanzees, but not with gorillas. Gorillas, humans, and chimps share an additional thirty-three Alu inserts not shared with orangutans.[15] The only way that pattern could happen is for seven different ancestors of all chimps and all humans each to have been "infected" with seven new Alu inserts. Finlay gives a long list of such viral insertions and pseudogenes shared by humans and chimps.[16] This data puts us humans (genetically speaking) right in there with the rest of them! The obvious biological explanation is that chimps are our first cousins. By way of comparison, the genetic distance between sun bears, brown bears, and black bears is about the same as the distance between chimps, gorillas, and humans. The implication is either that humans and chimps share a fairly recent common ancestor, or that God made human genes to look as if they were the genes of a third species of chimpanzee.

Humans' Genetic Differences from the Rest of Creation

There is, however, other genetic evidence that must be explored. Genetically speaking, the human species is very different from any of the ape species. First, we have very low local differentiation.[17] This means that as species go, local groups of humans are amazingly alike. Any given local human population—Oslo, Norway, or Beijing, China—has about 86 percent of the total genetic diversity of the entire human race (species). That is about the percent overlap that a local chimp population would have with its local *subspecies*. Consider that the chimps of Kenya and the chimps of Nigeria share only about 40 percent of their genetic traits. But human beings are more closely related genetically to one another across the globe than the chimps of Kenya are to each other. Compared to the apes, the human race does not have subspecies. This means that those characteristics that we think of as "racial" are completely superficial. We are truly unique in this regard.

Also, humans have far less overall genetic diversity than do any of the ape species—even the rare pigmy chimp (the bonobo). A single troop of bonobos typically has more genetic diversity than the entire human race. This has been measured in a whole series of genetic systems (mitochondrial DNA, for instance). The obvious biological explanation is that not so long ago, perhaps one hundred thousand years, our whole species was much smaller and more localized than the local populations of any of the living ape species have ever been. We went through a "genetic bottleneck." Our genetic divergences reach back only to that point. Chimps and bonobos, on the other hand, have retained a lot of genetic diversity, all the way back to their point of divergence around two million years ago.

When these genetic realities first were understood and reported in 1989, the impact on paleoanthropology was explosive.[18] The accepted view prior to 1989 ("regional continuity"—a view still held by many) was that local archaic hominid populations had developed in semi-isolation into local modern populations. If so, the genetic evidence would show far greater human genetic diversity than is found in any of the species of the great apes, since *Homo erectus* spread much farther than they did—from Africa to Europe and Asia about 1.5 million years ago. (Gene diversity is conserved in widely scattered populations—known as the Walund effect.) But the new genetic data suggested another model, one in which all human populations arose around 160,000 years ago from a single, small local population.[19]

Genetic Evidence and Human Origins

Since the first reports, literally hundreds of genetic studies of human populations have been done using a wide variety of genetic systems. The consensus of almost all of them is that our species was very small indeed around 160,000 years ago. Furthermore, the oldest populations, those with the most internal diversity, are those of Africa, possibly from around the Rift Valley. Growing from such a small ancestral population is termed a "genetic bottleneck." But is such an event biologically reasonable?

There is indeed a "smoking gun," clear evidence pointing to a necessary bottleneck. All primate species except humans have two chromosomes that are homologous to the two arms (halves) of our second chromosome. (We can tell that they are homologous by comparing the order of the DNA bases/genes on them.) Our chromosome 2 clearly was created by fusing two ape-like chromosomes. The fusion happened between the telomeres (the special end points) of the two short arms. The fusion chromosome that was produced had two widely separated centromeres (the mitotic control area of a chromosome). Relic telomeric DNA and centromeric DNA still exist at the spots on human chromosome 2 where these changes took place. That fusion chromosome, which was first formed in one specific individual, has become the only established human chromosome pattern.[20]

Any creature with a fusion chromosome like that would have had a profound reduction in fertility. The chromosomes usually would break during sperm or egg formation. Even when the extra centromere was deactivated, "crossbreeds" (individuals with one set of chromosomes of each pattern) still would have only about one-third normal fertility. In any large population natural selection would immediately get rid of such a deleterious fusion chromosome.

But the fusion type is the only one found in humans. It could replace the original chromosome only in a small, isolated population. Just by "chance," mostly "fusion babies" might be born (genetic drift), and then selection would weed out the ancestral (unfused) form within a very few generations. And then, when the fusion chromosome is all there is, everyone has full fertility—as long as they do not breed with the outside world (all the other hominids on earth). Bingo! We have a new species: human beings.

What this means is that all people on earth are descended from one individual who had that chromosome fusion. From a geneticist's perspective, this point is certain. In most mammals this sort of a strong fertility barrier will drive the quick establishment of "pre-reproductive isolating mechanisms"—perhaps the changes in face shape (the sharp chin, for instance) were to help the new humans recognize each other.[21]

Where did this probably happen? Since peripatric (on the periphery) speciation requires significant isolation, the middle of the African savanna seems unlikely. It should be close, however. The most likely place is an isolated spot on the edge of that African area (Ethiopia) where we see the first modern humans—perhaps to the northeast, in the fertile valley where the Persian Gulf is today. The new human species could then easily migrate to northeast Africa, spinning off the bands of ancient wanderers who worked their way down the endless savannas of eastern Africa to leave their bones at the Klaises River mouth in South Africa.

More Genetic Clues to Our History

Another interesting feature of these genetic studies is that they apparently can reveal pieces of the unique histories of specific human populations. We have read in news reports of their use in some specific cases—finding Thomas Jefferson's descendants, for instance. But they can reach much farther into the past. Genetic studies indicate that people migrated to Southeast Asia by eighty thousand years ago, beating the first colonists entering Europe by forty thousand years. And European genes indicate that when the first farmers moved there from the Middle East, they passed on their techniques much more rapidly than their genes.[22]

As for the Neanderthals, some of their DNA has been isolated (from eight fossil individuals)[23] and sequenced. Their genes were equally divergent from the genes of all modern human populations, different enough to suggest that we share a common ancestor with Neanderthals at around five hundred thousand years. They are a lot closer to us than chimpanzees are, but they are far off the mark for being the ancestors of modern Europeans. Apparently, they died out with no descendants, leaving Europe to modern immigrants from Africa.

The pieces seem to fit together to form a picture. The genes are telling the same story as the stones and bones. Around 160,000 years ago, a small African population of modern-looking people can be spotted on the shore of the Red Sea trying out new ways of doing things. They are our ancestors. They spread out across the globe, and all the scattered archaic populations of hominids died out as they arrived. That does not mean that we killed them, but the genes do tell us we did not breed with them.

Does this mean that Adam lived at that time? Can the biological data support the idea that modern humans are descended from a single couple? Was it Adam who had that fusion chromosome? If so, most of the genetic data would look very much like what we see.

Certainly, this picture is far closer to the biblical concept than it has been for a long time, although it still raises theological issues. But there is another biological problem. Some genetic systems imply that there were always more than two people in the lineage, perhaps twenty or so as an absolute minimum. (HLA antigen proteins—part of our immune system—exist in many alternative forms that are so different that they could not have developed from just four versions, two each from Adam and Eve.[24]) This suggests that a small group of nonhuman creatures became isolated, and it was through this group, or someone in it, that God chose to establish the human race. It was at this point when the human race was created, that one of God's creatures first bore God's image. Further research may change this picture. This is an area of active, and contentious, research. But for now, that is what the data look like.

Another issue is that the biblical story is told in the cultural context of early agriculture—the Bronze Age. That culture dates to around 10,000 to 6,000 B.C. Ancestors of modern populations were all over the world by then. This is no real problem, since the storytellers themselves are of the Bronze Age—that is, of course, the only reference from which they could interpret or tell of events from a far earlier milieu. Tubal-Cain, for instance—perhaps the "father of all those who work metal" (see Genesis 4:22) is the memory of an inventive Aurigancian flint-knapper.

But by how much did human beings predate the accounts of the first human beings in Scripture? (This would affect how we interpret Genesis, particularly the creation account.) What about the archaic hominids? Were they human? Perhaps we can never be sure, but given the cultural evidence that we have (or rather, what we do not see, such as art), they probably were not human.

They seem much like us in some ways, but if the image is a commission from God, not just a description of characteristics such as brain size, evidence of that commission seems to be missing in archaic hominids.

But could certain archaic populations be our ancestors? There are indeed evidences that some archaic populations in North Africa were getting close to modern morphology around two hundred thousand years ago. Could a small population of these beings have been used to make the first real humans? Could they have been the clay molded by the Creator's hand into which God breathed his image? Or did God simply use them and their genes as a model for a new creation? He could, of course, do either. The results possibly could look the same. Direct creation indeed might be made to look like ancestry, but if so, God included the retrovirus inclusions from the chimpanzee in the package. Would God try to trick us?

Conclusions

What is a Christian to make of all this? Certainly, people who want to challenge the Christian faith use this information, as they use all information, to attack. Quite possibly, when you have heard information about the physical origins of the universe or of human beings, the context has been one of challenging Christian beliefs and principles. But the data, and even the theories, have no necessary threat in them. It is true that they do not sound much like the traditional understanding of human creation. But then, this is a description of the *outward appearance* of the creation, whereas the Scriptures give us a description of the inside story, *the meaning*.

It is God who is molding the stuff of earth on his potter's wheel to make human beings. We are dust, the Scriptures tell us, made of the stuff of earth. So are the flesh and the genes of apes. And so is the mud in a creek bed. How long the wheel spun, and how much time God took to mold us, we might interpret as evidence of how careful God was, just as we interpret the incarnation as part of

God's effort to redeem humanity, even though he might have been able to do this in some other way.

As we have noted time and time again, God could have created the whole universe instantaneously. We should not ignore evidence that God chose to do it differently. Certainly, God breathing spiritual life into Adam addresses something other than physical life (after all, mosquitoes are alive too), and certainly it is not an event that we can expect to see in the fossil record.

What, then, should we think of Adam and Eve, the serpent and the garden, the fall and the curse? These are theological issues. There are many different interpretations of the stories in these passages of Scripture. Some clearly would be very difficult to correlate with the aforementioned "scientific" data, while some would fit comfortably with it. The scientific data may suggest that theologians need to rethink their models of what the Scriptures are saying, but it does not tell them what the correct models should be.

I am not a theologian. However, since I believe that God is truth in all he does, I tend to accept the interpretation that fits *both* God's Word and God's world. But hold your interpretations lightly. Science and theology, being human activities, change with time— unlike the inspired Word and the created world, both of which are anchored in the eternal God of truth.

CHAPTER 14

Evolution as Creation

L ET US BEGIN WITH A RECAP OF WHAT WE HAVE ACKNOWLEDGED and uncovered. Some of what appears in this concluding chapter the reader will recognize from previous chapters, but I repeat it here for emphasis.

We begin with the understanding that questions of science are answered by studying the data. And in sum, the data tell us that all living things are united in their use of the same "language," the same genetic code. Mutated genes produce changed proteins, and changed proteins produce changed bodies. Genetic "fossils" are linked with living species in genealogical trees of descent.

New species can be observed forming in the present and recent past, the polar bear being one example. The fossil record contains hundreds of examples of organisms forming a bridge between existing and extinct types. Indeed, some whales walked with legs, and some dinosaurs sported feathers.

Statements like these are just as true of human beings as they are of other species. These are not the *theories* of science; they are reliable *descriptions* based on observation of the data.

No doubt, some people have already decided that creation as described in the Bible and the evolutionary processes described in this book are incompatible. I recognize that there is likely to be nothing anyone can say to change this thinking. Such persons would need to write me and many others off as being either deceitful Christians or incompetent scientists. The same dynamic was in place a few centuries ago when many in the church simply could not conceive of a God who would create a universe in which the earth was not the center. When Galileo asked his critics to look though his telescope and see sunspots, lunar mountains, and the moons of Jupiter—data that supported the theory of Copernicus—they refused to look because it might "twist their minds." But this did not erase the data, and it did not prevent these critics from being wrong.

All I can ask is that you keep an open mind, that you remember any *theory* of origins must include all these pieces, just as any case in court must include all the evidence. Collectively, the facts provide virtually undeniable evidence for an evolutionary process. But not one of these facts even hints that God was not involved.

Science has as its primary responsibility explaining the world of nature, and theology has the primary responsibility of explaining the nature of God. However, for the Christian, the Scriptures tell us, nature is the world God created. This means that a Christian doing science must consider the words of Scripture, and a Christian doing theology must consider the data of *created* nature. These two "books" of God—Scripture and nature—cannot be in conflict.

As we learn more about what God has done in nature—that is, as we do science—we should be prepared to find that some of what we thought Scripture meant about nature (not about sin and salvation) needs to be revised. Discussing the concept of "day" in Genesis, Charles Hodge, the very conservative nineteenth-century Princeton theologian, stated the point this way:

> It is of course admitted that taking this account by itself, it would be most natural to understand the word in its ordinary sense; but if that sense brings the Mosaic account into conflict with fact and another sense avoids such conflict, then it is obligatory on us to adopt that other....As the Bible is of God,

it is certain that there can be no conflict between the teachings of the Scriptures and the facts of science. It is not with facts but with theories, believers have to contend....The church has been forced more than once to alter her interpretation of the Bible to accommodate the discoveries of science. But this has been done without doing any violence to the Scriptures or in any degree impairing their authority. Such change, however, cannot be effected without a struggle.[1]

Unfortunately, many theologians and pious Christians have since concluded that such revision constitutes unfaithfulness to the Scriptures. The implication is that people of faith should ignore the evidence of the senses and the discoveries of science in favor of taking the Bible "literally." Sadly, this position has done great damage to the mission of the church. How can you persuade people to evaluate the claims of Christ when you deny obvious truths of which they are certain?

A Time of War

It is true enough that the faith is under attack. And it is true that this attack has frequently and unjustifiably claimed the mantle of science. This is nothing new. Throughout its history, the church has been attacked by those who are committed by faith to the absence of God. But it is not science that is doing the attacking. There are, and have been since the nineteenth century, vocal advocates of the view that science has invalidated any rational grounds for belief in God. Since Darwin's day, the creative power of "evolution" has been among their staple arguments. But, again, this is the voice of atheism, not the voice of science. As we have seen, there is no *necessary* tie between the processes of evolution and atheism.

In a time of attack, of course, it is important to defend the faith. However, in a shooting war, many a soldier dies from "friendly fire." In the conflict that is the topic of this book, "friendly fire" from well-meaning Christians can steal souls just as surely as does "enemy fire" from the atheist camp. Sunday schools and earnest preaching, if what they teach is not truthful, can easily drive people away from faith rather than drawing them to it.

Consider a specific example. Imagine a preacher who tells a medical student that the Bible teaches that men have one fewer pair of ribs than women have. Then the preacher says with equal certainty that Christ is the incarnate son of God. The medical student dissects a cadaver and determines that what the preacher said about the number of ribs that men have is untrue. The student likely will question the credibility of anything this preacher has to say. If the speaker is obviously ignorant, the student might just write it off as general ignorance about everything, including the Bible. However, if the speaker is a person who is theologically educated, the student is likely to take his or her word for what the Bible says and therefore reject the Bible as a book of errors. The student might never again consider the claims of Christ. As the eminent theologian Augustine (A.D. 354–430) put it,

> Usually, even a non-Christian knows something about the earth, the heavens, and the other elements of this world, about the motion and orbit of the stars and even their size and relative positions, about the predictable eclipses of the sun and moon, the cycles of the years and seasons, about the kinds of animals, shrubs, stones, and so forth, and this knowledge he holds to as being certain from reason and experience. Now, it is a disgraceful and dangerous thing for an infidel to hear a Christian, presumably giving the meaning of Holy Scripture, talking nonsense on these topics; and we should take all means to prevent such an embarrassing situation, in which people show up vast ignorance in a Christian and laugh it to scorn. The shame is not so much that an ignorant individual is derided, but that people outside the household of the faith

think our sacred writers held such opinions, and, to the great loss of those for whose salvation we toil, the writers of our Scripture are criticized and rejected as unlearned men.... Reckless and incompetent expounders of Holy Scripture bring untold trouble and sorrow on their wiser brethren when they are caught in one of their mischievous false opinions and are taken to task by these who are not bound by the authority of our sacred books. For then, to defend their utterly foolish and obviously untrue statements, they will try to call upon Holy Scripture for proof and even recite from memory many passages which they think support their position, although they understand neither what they say nor the things about which they make assertion.[2]

Unfortunately, it is not at all difficult to find theologians who fit Augustine's description. Although they intend to defend the faith, they cannot be bothered with the facts, the data of science. Thus they attack what they do not understand. They teach errors of fact, and they slander fellow Christians who are trying to put the pieces together. In so doing, they make Christian faith a laughingstock, and—make no mistake about it—atheists love them.

Many times, Christian graduate students in the sciences have told me of their experiences in witnessing. Their fellow students are inclined toward faith, sometimes because of the wonderful ways in which they see nature operating. But ultimately they turn away because they think they have to accept ideas such as the "young earth" to be a Christian, and they cannot commit intellectual suicide. They cannot choose to believe something they know to be false. No one can.

Walking by Faith?
The preceding chapters have discussed a wide variety of controversial ideas. Where do Christians go from here? How should we

handle evidence for evolution? Many Christians simply say that we should not even listen to such ideas, that a real believer would simply rest on "faith." Such persons urge us to "have faith in God," but actually they are urging us to have faith in them—that is, to accept the position (in scientific terms, the theory) that they are advocating. After all, strong faith believes regardless of the evidence, does it not? Indeed it can. After all, what allows con artists to succeed? Strong beliefs based on falsehoods.

Theories (whether in science, police work, or theology) may indeed be wrong, but again, the data simply are what they are. The data cannot be wished away when they do not fit neatly into our accepted view of reality.

How can we as persons of faith be honest, humble, and objective in our search for truth? We can begin by acknowledging that we operate out of a worldview that enables us to see patterns and to find purpose. In fact, both Christianity and materialism are comprehensive worldviews. As with all worldviews, they shape (bias) our perceptions of data and patterns. If we do not admit that we have such personal faith commitments (personal biases), we will persuade ourselves that our viewpoint is the only rational view.

Beyond that, we can try to do the following. When the data seem to confirm our commitments, seriously seek other possible interpretations. Be slow to reject interpretations that we dislike. Wait for further evidence. "Play the devil's advocate" (before someone else does). We do not want to find ourselves embracing easy falsehoods in the service of "truth." Although our children may find it convincing today if all our answers are nailed down, tomorrow we will lose the children if the weight of neglected or rejected data pulls out our nails. Those who perceive our falsehoods will also reject our truths.

Science consists of the formation of theories as part of the effort to explain the data. For the Christian in science, theory formation and testing is akin to wrestling with an angel. Our commitment to truth is a commitment both to the truth of God's Word and to the data of God's creation. Both our worldview commitment and the shape of the reality that meets our senses force themselves upon our theoretical struggles. There are nonnegotiable pieces at both ends of Jacob's ladder. But remember that the materialist is no better off, no more objective, unbiased, or neutral.

Tagging the Enemy

The real war has been one of long standing. And it is not about science. Scientific theories usually do not inspire a fight to the death. Rather, our religious commitments, our fear of change, our desire for power and prestige—these bring war, with gallows and guns or with derogatory words and smearing associations. But groups that hold (or wish to overthrow) established positions of power and prestige will fight using whatever weapons they can find.

Times change. The conflict between the naturalism movement and the old establishment (where the church held a place of privilege) has been replaced by the conflict between the creationist movement and the new establishment (where the academy holds a place of privilege).

Consider the following conundrum: Almost none of the membership of the National Academy of Science will admit to a belief in a God who answers prayer, yet around 40 percent of the "rank and file" (members of *Who's Who in American Science*) do so believe.[3] Is materialism a loyalty test for the present "establishment," assuring that the prestige of science remains the banner and possession of the inheritors of the naturalism movement? It seems likely. But why would they care?

Why do we humans set up loyalty tests? It is because in war, you must be able to spot the enemy. Clarity and certainty are critically important. Both sides in a battle shoot at anyone in "no man's land," and both sides identify any dissent or disagreement as treason. "Beloved" enemies are those who prove my prejudices correct! They strengthen my resolve. "Hated" friends raise issues that we wish would be taken for granted. They weaken my resolve. But what does that have to do with the question of evolution? Simply this: The term "evolution" has become the touchstone to divide the camps of "science" and "Christianity." Under these conditions, even *calling* for an objective discussion is considered both treacherous and blasphemous.

Is evolution a threat to the Christian faith? Most emphatically, no! The real threat is the materialistic faith of the naturalism movement. If evolution means embracing the faith of the materialists, then of course evolution is, by definition, a threat. If evolution simply means genetic relationships and/or natural forces, then the larger meaning and significance of evolution depend on the worldview of those using the word. Do the persons in question view nature as autonomous or as governed by God? For those who view nature as governed by God, evolution is no threat. It is, in fact, the hand of God at work in his world. Why should we as Christians allow naturalism to define the word?

But this is not how many people see it. In fact, there are three assumptions that keep the word "evolution" at the center of the fight, assumptions shared by those on both sides of the debate. First, they view the activities of nature as autonomous, independent. In this view, if God exists and acts, God acts only by setting initial physical conditions, or by resetting those conditions. After such points of input, the system proceeds by itself. Second, their model for understanding God as creator is the human engineer and nothing more. Thus, creation is best understood exclusively in terms of engineering efficiencies. Likewise, "God the engineer's" purposes are limited mostly to the outcomes or products of the creative process. Third, they feel that the Scriptures contain scientific statements of a sort, and that these statements are necessarily incompatible with any meaning of evolution (including common descent or change in body type).

These assumptions are made by a surprising mix of mutually

antagonistic groups. Young-earth creationists, intelligent-design theorists, materialists, advocates of the Blind Watchmaker—all make these assumptions as they argue. These groups may detest each other, but they also depend on each other. Hard-nosed advocates of atheism pushing the Blind Watchmaker hypothesis religiously empower the creationists. Creationists quick to reject the data of science surrender the scientific high ground to the materialists. Beloved enemies indeed.

But what does God think about these propositions? What does God's natural world say? What is the position of the Scriptures?

Interpretation and Literalism

This statement is frequently made: "I do not interpret the Bible; I just read it. I take the Bible for what it says." This sounds humble, but it is the humility of the Pharisee. For by refusing discussion, the speaker assumes the place of God. What this type of reader sees in a given Bible passage is nonnegotiable. His or her particular understanding is infallible; to disagree with that understanding is to disagree with God. Such manipulative statements frequently are used to push various theological positions, including creationism.

This approach has had a deadly effect on what the gospel is about. For, again, the data of nature are not negotiable. They are part of God's truth. To insist on giving a position on the meaning of the Word of God that is contradicted by the clear evidence of nature (and I do not mean the theories of science, but *data* such as 4,600 feet of coral) is to gut Scripture's authority. It is, in fact, to place the authority of the interpreter above the truth of God. Frequently, it serves primarily as a wedge to give the interpreter power and prestige.

Of course, most people who make such statements about the Bible do not mean to be arrogant. Many people feel ill equipped to address the specifics of the controversy at anything more than a surface level. What they may be saying is, "I accept the meaning given to this passage by people I trust, and *they* are taking the Scripture for what it says." They usually have a naïve misunderstanding of how we know things. They assume that everyone sees things "the way they are." But that view has long since been proved inadequate. Everything that we see we interpret though the glasses of our worldview—our culture as filtered through our personal experiences.

As historian Mark Noll puts it,

> When evangelicals rely on a naïve Baconianism, they align themselves with the worst features of the naïve positivism that lingers among some of those who worship at the shrine of modern science. Thus, under the illusion of fostering a Baconian approach to Scripture, creationists seek to convince their audience that they are merely contemplating simple conclusions from the Bible, when they are really contemplating conclusions from the Bible shaped by their preunderstandings of how the Bible should be read.[4]

How shall we understand the Scriptures? "Literally"? What does that mean? There are many models, but all are pointless if we do not realize that ultimately all readings of Scripture (and of nature, for that matter) are interpretations. What we call the "literal" meaning of the text is what we see at first glance through the eyes of our culture. But the "literal" meaning of a text is what it meant to the original hearers as they listened through ears trained by their culture. And they may not have heard what we heard nor concluded what we conclude. The standard of biblical interpretation is to understand the passages as the original hearers would have heard them.[5]

This does not remove the Scriptures' authority; rather, it allows them to speak to us. If the Scriptures cannot be "translated" into the circumstances of our lives, if they must be taken "literally," they would not be able to speak to our realities. And the first step in this translation process is to understand the original meaning for the original audience.

Consider this example. Very few of us wash each other's feet at church, even though the Bible literally commands us to do that (see John 13:14). We ignore this clear scriptural injunction because it is culturally specific. Modern Western culture, unlike that of the first Christians, does not do the act of footwashing. But if we understand the meaning of the act in their culture—hospitality and humility—we will engage in humble acts of hospitality suitable for our own culture. Is there a new family that you wish had not visited your church? Perhaps they were not wearing the right clothes, or they came from the wrong part of town. The Bible instructs us to "wash their feet," which for us might translate into inviting the family home for dinner.

Christians in the past have used literal interpretations to justify evil. We need look no further than the "biblical" justification made for the practice of slavery in recent centuries. And then there are the literal interpretations that we ignore because our culture condones them. For instance, various passages of Scripture place restrictions on lending money for interest, and in some cases forbid it altogether. In our culture, awash in mortgages, credit cards, and student loans, we assume that such Scripture passages are figurative statements and metaphors. Surely, we think, they do not apply to us.

The question of literalism certainly comes up when we read the Scriptures for insights into nature. What instinctive assumptions do we moderns make when we read descriptions of nature written by Middle Eastern herders and farmers of thousands of years ago? We in the Western world live in a world of mechanisms—cars, trains, airplanes, computers, factories full of machinery. We have no choice. We were born into it. We are far, far removed from animistic culture out of which the Scriptures were born.

Thus we interpret the biblical writers' statements about the way

God governs and cares for creation as being merely figurative and poetic. We read their statements as if we had written them, considering only what we would have meant by such "poetic" utterances. And when the Biblical writers speak poetically/symbolically, we often look for a physical meaning. And then we call it "literal"!

When the biblical writers refer to wind and fire as God's ministering servants—"He makes winds his messengers, flames of fire his servants" (Psalm 104:4)—they meant it literally. They worshiped a God who was actively at work, present in the world. In our world of machines we have by and large abandoned this perception of God. To the extent we have done so, we have become functional deists.

Consulting King David

What we need is an ancient local native to translate what Genesis would have meant to him. And perhaps, with a bit of imagination, we can do this. I suggest what follows as King David's impressions of the creation, a parallel to Genesis 1 written down in Psalm 104. First, here is the psalm (KJV):

> Bless the LORD, O my soul. O LORD my God, thou art very great; thou art clothed with honour and majesty. Who coverest thyself with light as with a garment: who stretchest out the heavens like a curtain: Who layeth the beams of his chambers in the waters: who maketh the clouds his chariot: who walketh upon the wings of the wind: Who maketh his angels spirits; his ministers a flaming fire: Who laid the foundations of the earth, that it should not be removed for ever. Thou coveredst it with the deep as with a garment: the waters stood above the mountains. At thy rebuke they fled; at the voice of thy thunder they hasted away. They go up by the mountains; they go down by the valleys unto the place which thou hast founded for them. Thou hast set a bound that they may not

pass over; that they turn not again to cover the earth. He sendeth the springs into the valleys, which run among the hills. They give drink to every beast of the field: the wild asses quench their thirst. By them shall the fowls of the heaven have their habitation, which sing among the branches. He watereth the hills from his chambers: the earth is satisfied with the fruit of thy works. He causeth the grass to grow for the cattle, and herb for the service of man: that he may bring forth food out of the earth; And wine that maketh glad the heart of man, and oil to make his face to shine, and bread which strengtheneth man's heart. The trees of the LORD are full of sap; the cedars of Lebanon, which he hath planted; Where the birds make their nests: as for the stork, the fir trees are her house. The high hills are a refuge for the wild goats; and the rocks for the conies. He appointed the moon for seasons: the sun knoweth his going down. Thou makest darkness, and it is night: wherein all the beasts of the forest do creep forth. The young lions roar after their prey, and seek their meat from God. The sun ariseth, they gather themselves together, and lay them down in their dens. Man goeth forth unto his work and to his labour until the evening. O LORD, how manifold are thy works! in wisdom hast thou made them all: the earth is full of thy riches. So is this great and wide sea, wherein are things creeping innumerable, both small and great beasts. There go the ships: there is that leviathan, whom thou hast made to play therein. These wait all upon thee; that thou mayest give them their meat in due season. That thou givest them they gather: thou openest thine hand, they are filled with good. Thou hidest thy face, they are troubled: thou takest away their breath, they die, and return to their dust. Thou sendest forth thy spirit, they are created: and thou renewest the face of the earth. The glory of the LORD shall endure for ever: the LORD shall rejoice in his works. He looketh on the earth, and it trembleth: he

toucheth the hills, and they smoke. I will sing unto the LORD
as long as I live: I will sing praise to my God while I have my
being. My meditation of him shall be sweet: I will be glad in
the LORD. Let the sinners be consumed out of the earth, and
let the wicked be no more. Bless thou the LORD, O my soul.
Praise ye the LORD.

The psalm begins with the glorious King in his palace. It speaks
of the forces of nature, making it clear that these forces act as his
obedient servants, carrying out his commands. Each part of his
realm (sky, earth, and seas) is established, governed, shaped, and
ruled by his word of command. Furthermore, the King oversees the
supplying of the needs to the inhabitants of the kingdom, whether
beasts, birds, whales, or people. In return, each inhabitant has its
own role, its obligation to its sovereign. Not an absentee sovereign,
the King actively rules over the creatures, dispensing life and death,
destruction and creation (*bara*). This vivid picture shows us that
David did not see the creation as a machine (as we do), or as an
organism (as the Greeks did), but as a kingdom with an absolute
Eastern monarch.

What, then, did the disciples hear when Christ said, "Not a single sparrow falls except by the will of my Father"? Not just a beautiful poetic statement about God's care for them, but a description
of nature as they understood it to be. That is what "literal" should
mean: what it meant to *them*. John Calvin made the comment that
we cannot understand the role of God in creation until we understand God's role in providence. One can see the reason in this
psalm, for the Creator God is the reigning, providential King.

How did the psalmist hear in the first chapters of Genesis? "God
said, 'Let there be....'" What David heard was the King command-
ing his servants to "make it so." In Psalm 104 God's servants, the
natural forces of light, fire, wind, and water acted—"And it was so."
"And God saw that it was good." In this last statement, our guide

heard the King's judgment of the work of his servants. Likewise, the commands came to the ground, to the waters, to the cosmos. "Let the land produce vegetation....And it was so. The land produced vegetation." And so our guide again would have seen the physical world obey, producing life, empowered by its King. And again, "And God saw that it was good." And so the command went next to the living, moving creatures (echoed in Psalm 104), telling them to reproduce and fill the earth. Furthermore, God established their provisions. They were clients of his grace, as our psalmist guide noted (in Psalm 104), obedient and dependent.

The psalmist then noted that humanity was also made for the kingdom and commissioned to high office under the King, imaging God in the creation. But yet, though molded by their King like a potter, humanity is still part of the kingdom, dependent on the King's provision, like all the other creatures. Our guide would have assumed the obvious, "literal" meaning of dust in his culture. That literal meaning would have had nothing to do with the physical nature of soil. Rather, dust referred to humble origins, to the idea that human beings are servants of the King, raised up from the dust and returning to it when the King's favor is withdrawn.

There is another lesson that our psalmist guide would have heard in Genesis. None of the servants of the King (forces, creatures, or humans) had divine (or kingly) authority. All their powers were bestowed by their King. The surrounding peoples viewed natural forces as independent—in fact, as having divine creative powers (the Baals). But the psalmist knew that each kind of creature was completely dependent on, governed by, and answerable to the one true King.

So what would David think about the evolution controversy? Certainly, he would not fit easily into the standard battle armor. Clearly, he would not hold the first assumption supporting the conflict. His world was a governed kingdom, not an autonomous machine. Likewise, he would reject the second assumption. His God's creation was a function of kingship, not an engineering fabrication.

As for scientific truth in the Scripture, the psalmist would need a bit of education in what scientific "statements" are. However, the young man who killed bears and lions in defense of his flock must have had a fairly good working knowledge of biological reality. In Psalm 104 he attributes the forces involved in the evolutionary mechanisms—lions feeding, birds nesting, deaths and rebirths—to the direct continuing actions of the Creator King. He would not have been tempted to view such "forces" as autonomous. He would indeed have identified the idea of free-acting nature as Baal worship.

The psalmist who wrote, "The heavens declare the glory of God; and the firmament sheweth his handiwork. Day unto day uttereth speech, and night unto night sheweth knowledge" (Psalm 19:1-2, KJV), would have taken natural data as revealing God's glory. Deep space and deep time as we know them would show only how immense, how far beyond us, God is ("With the Lord one day is like a thousand years, and a thousand years are like one day" [2 Peter 3:8; cf. Psalm 90:4]).

The vision of the infinite Creator shaping a biological world of unthinkable complexity in a universe vast beyond all human thought, working across and through unknowable ages of time, certainly is compatible with the vision of the psalmist. He would rejoice in the glory of God so revealed. And so should we.

If we step back from the shooting and shouting of the war between evolution and creation, we can gain a vision of the greatness of God that no generation before us could have had. God is so much greater than we knew! Our vision of God was so limited, our God was so small, because our knowledge of what God had made was so limited. When we discover that the creation is vast beyond our knowing, we realize that the creation is bigger than our vision of God. Some choose to reject their little God as inadequate;

some choose to reject the expanded vision of the creation that modern science has revealed. But the action of humble faith is to expand our vision of our God and to fall on our knees before God's infinite glory.

Evolution as Creation: The Lord of the Dance

Is the biological world indeed being led in the dance? How can this creature—this awe-inspiring world—have become so creative? Does this creativity reflect the character of God? Of course it does. Is the kingdom of God created as a mustard seed or a full-grown sequoia? We see God's creativity unfolding before us in the natural world. A creative creature certainly would not reflect the devil's character! That proud spirit would never give pride of place to that which he ruled.

Have materialists mistaken the creature for the Creator because the creation reflects "his eternal power and Godhead"? This view brings to mind a cut flower. How would the living world be able to sustain its creativity without its connections to its Creator? Why would it not wither away? What sort of structure would different universes have? What does creation look like to a God whose most notable characteristic is love?

One of the reasons that the evolutionary scenario appeals to many is the sense of "totality," of all reality being tied into one story. But so is creation reality tied into one story. And the vast sweep of evolutionary development in the world of life is part of that story of creation. One story indeed, but we may have missed the point of the plot! It is not about us. The world was not made simply to lead up to humanity. As God's unique image bearers, we are part of the script, but the plot of the world is to show forth the glory of its Creator.

I do not believe in God because I can prove God through science any more than I need proof to believe in the love of my wife. I believe in both on the basis of all the experiences of my life.

Certainly, I would be a fool to keep on believing if the existence of God was scientifically proved to be untrue, but by the very nature of the "God hypothesis," that cannot be done. Indeed, the patterns of data that we have examined in this book show that all such proofs depend on infinite knowledge of the probability of what has occurred. But for my money, given the data, it takes a lot bigger leap of faith to be sure that God is absent than to be sure that God is present.

Both theists and materialists have a "God hypothesis." Christians believe in the God of the Bible, the author of nature, while materialists hold that nature *is* God. In a battle between two true believers, neither can win over the other by argument. There is no resolution. They do not live in the same world. They do not swim in the same water. They cannot see the same ultimate patterns in the data. One calls evolution a dance led by the sovereign ruler of creation, while the other calls it "a tale told by an idiot, full of sound and fury, signifying nothing." Neither will be convinced unless and until the reality behind the patterns is revealed. And I am convinced that the reality behind the whale with legs and the feathered dinosaur is not the blind face of an impersonal, unthinking universe, but Yahweh, the Creator God of Israel.

Notes

Chapter 1

1. John Calvin: *Institutes of the Christian Religion*, edited by J. T. McNeill, trans. F. L. Battles (Philadelphia: The Westminster Press, 1960), Vol. 1, Chapter 16, 197-8.
2. From "The Confession of Faith," in the Constitution of the Synod of the Reformed Presbyterian Church, 1949.
3. Taken from comments about Aslan in C. S. Lewis's novel *Prince Caspian.*

Chapter 2

1. There is a considerable amount of technical literature on this effect, but an interesting popular and practical discussion of the implications is an article by psychologist Robert Adler, "Pigeonholed," *New Scientist,* September 30, 2000, 39–41.
2. The concepts of paradigm shift and scientific revolution are closely associated with the work of philosopher Thomas Kuhn, *The Structure of Scientific Revolutions* (Chicago: University of Chicago Press, 1970).
3. For further discussion of the nature of science, I recommend a book by philosopher Del Ratzsch of Calvin College, *Science and Its Limits: The Natural Sciences in Christian Perspective,* 2nd ed. (Downers Grove, Ill.: InterVarsity Press, 2000).

Chapter 3

1. Of the many excellent books that discuss the historical relationship between faith and science, here I recommend two: Colin Russell, *Cross-Currents: Interactions between Faith and Science* (Grand Rapids: Eerdmans, 1985); David Lindberg and Ronald Numbers, eds., *God and Nature: Historical Essays on the Encounter between Christianity and Science* (Berkeley: University of California Press, 1986).
2. J. R. Moore, "1859 and All That: Remaking the Story of Evolution and Religion," in *Charles Darwin, 1809–1882: A Centennial Commemorative,* ed. R. G. Chapman and C. T. Duval (Wellington, New Zealand: Nova Pacifica, 1982), 167–94.
3. E. J. Larson and L. Witham, "Scientists Are Still Keeping the Faith," *Nature* 386 (1997): 435–36.
4. Moore, "1859 and All That."
5. A long list of the books written by these individuals could be given. Probably the best known is Richard Dawkins, *The Blind Watchmaker: Why the Evidence of Evolution Reveals a Universe without Design* (New York: Norton, 1987).
6. Davis Young, a geologist at Calvin College, has written extensively on the response of the church to old-earth geology. Two of his books are *Christianity and the Age of the Earth* (Grand Rapids: Zondervan, 1982) and *The Biblical Flood: A Case Study of the Church's Response to Extrabiblical Evidence*

(Grand Rapids: Eerdmans, 1995).

7. Here is a sampling of books that evaluate the controversy surrounding the rise of creationism: A. Desmond, *The Politics of Evolution* (Chicago: University of Chicago Press, 1989); E. J. Larson, *Summer for the Gods: The Scopes Trial and America's Continuing Debate over Science and Religion* (New York: Basic Books, 1997); G. M. Marsden, *Fundamentalism and American Culture* (Oxford: Oxford University Press, 1980); R. L. Numbers, *The Creationists: The Evolution of Scientific Creationism* (New York: Knopf, 1992).

8. Larson, *Summer for the Gods.*

9. Numbers, *The Creationists.*

Chapter 4

1. Jeffery Greenberg, a geologist at Wheaton College, provides an excellent summary of the evidence for an old earth: "Geological Framework of an Evolving Creation," in *Perspectives on an Evolving Creation,* ed. K. B. Miller (Grand Rapids: Eerdmans, 2003).

2. The data about the coral reef is from Daniel Wonderly, *God's Time-Records in Ancient Sediments* (Flint, Mich.: Crystal Press, 1977). Wonderly also authored a volume entitled *Neglect of Geological Data: Sedimentary Strata Compared with Young-Earth Creationist Writings* (Hatfield, Pa.: Interdisciplinary Biblical Research Institute, 1987).

3. In addition to the previously cited works by Davis Young (see note 6 for chapter 3), Jeffery Greenberg, and Daniel Wonderly (see notes 1 and 2 here), I recommend H. J. Van Till, D. A. Young, and C. Menninga, *Science Held Hostage: What's Wrong with Creation Science and Evolutionism* (Downers Grove, Ill.: InterVarsity Press, 1988). The reader will have noticed that all of these references are by Christian scientists and from Christian publishers. Of course, a secular text will present the data for an old earth as well, but not as an argument—rather, as established truth. One example is Steven Stanley, *Earth System History* (New York: Freeman, 1999).

4. For more details on the Cambrian era, I recommend Simon Conway Morris, *The Crucible of Creation: The Burgess Shale and the Rise of Animals* (Oxford: Oxford University Press, 1998).

5. Greenberg, "Geological Framework."

6. Colin Russell, *Cross-Currents: Interactions between Faith and Science* (Grand Rapids: Eerdmans, 1985); David Lindberg and Ronald Numbers, eds., *God and Nature: Historical Essays on the Encounter between Christianity and Science* (Berkeley: University of California Press, 1986).

7. Davis Young, *Christianity and the Age of the Earth* (Grand Rapids: Zondervan, 1982).

8. R. L. Numbers, *The Creationists: The Evolution of Scientific Creationism*

(New York: Knopf, 1992).

9. Ibid.

10. J.C. Whitcomb and H.M. Morris *The Genesis Flood: The Biblical Record and Its Scientific Implications* (Philadelphia: Presbyterian and Reformed Publishing Co., 1961)

Chapter 5

1. Retroviruses are a group of viruses that can put copies of themselves into a host chromosome and then be released again. Sometimes they take a gene from the host along for the ride, infect a new host, and insert the gene from the first host into the new host's chromosome. This process is termed the "lateral transfer" of a gene.

2. As Richard Dawkins uses the expression "blind watchmaker," the material universe is the purposeless autonomous creator. See Richard Dawkins, *The Blind Watchmaker: Why the Evidence of Evolution Reveals a Universe without Design* (New York: Norton, 1987).

3. Two examples of the nineteenth-century writers are Lyman Abbott and E. D. Cope. Abbot was a preacher, Cope a paleontologist. Both identified evolution with an intrinsic force for progress. See Lyman Abbott, *The Theology of an Evolutionist* (Boston: Houghton, Mifflin, 1897); E. D. Cope, *The Origin of the Fittest: Essays on Evolution* (New York: Appleton, 1887).

4. Teilhard de Chardin, *The Phenomenon of Man* (London: Wm Collins, 1959).

5. Theodosius Dobzhansky, "Nothing in Biology Makes Sense Except in the Light of Evolution," *The American Biology Teacher* 35 (1973): 125–29.

Chapter 6

1. O. Rieppel, "Structuralism, Functionalism, and the Four Aristotelian Causes," *Journal of the History of Biology* 23, no. 2 (1990): 291–320.

2. B. B. Warfield, "A Review of *Darwinianism Today,* by Vernon L. Kellogg," *Princeton Theological Review* (1908): 640–50.

3. G. G. Simpson, *The Meaning of Evolution: A Study of the History of Life and of Its Significance for Man,* rev. ed. (New Haven: Yale University Press, 1967).

4. The term "deism" is well known, usually being applied to the Enlightenment view of an impersonal clockmaker-creator who, having acted initially, vanished. Science historian Reijer Hooykaas noted that many Christians also hold this autonomous deistic materialism of the Enlightenment but allow the Creator to intrude upon occasion—to do a miracle. He termed that view "semideism." See Reijer Hooykaas, *Religion and the Rise of Modern Science* (Edinburgh: Scottish Academic Press, 1972).

5. A considerable amount of literature has been generated by the debate over intelligent design. I will consider the arguments in the next section, but here I list works by two of the better-known authors: P. E. Johnson, *Darwin on Trial* (Downers Grove, Ill.: InterVarsity Press, 1991); M. Behe, *Darwin's Black Box: The Biochemical Challenge to Evolution* (New York: Simon & Schuster, 1996).

6. Behe, *Darwin's Black Box.*

Chapter 7

1. One of the clearest summaries of this discussion is Simon Conway Morris, *Life's Solution: Inevitable Humans in a Lonely Universe* (Cambridge: Cambridge University Press, 2003). This book is an interesting argument for purpose in human origins, made by an interesting author, a well-known evolutionary scientist who is also a Christian.

2. The basic genetics being outlined here could be gleaned from a multitude of texts and popular presentations. For an amusing basic description see Martin Brookes, *Get a Grip on Genetics* (New York: Barnes & Noble Books, 2003).

3. Reviewing the sweep of existing theory is far beyond the scope of this book. Ranging from self-replicating clays (Cairn-Smith) to silicon intelligences (Fred Hoyle), from superheated undersea vents to drying mud flats to extrasolar civilizations (Francis Crick) the "holy grail" is a replicator prescribing a metabolic machine (Orgel). For further reading see R. Shapiro, *Origins: A Skeptic's Guide to the Creation of Life on Earth* (New York: Bantam Books, 1987); M. A. Edey and D. C. Johanson, *Blueprints: Solving the Mystery of Evolution* (Oxford: Oxford University Press, 1990), chapter 16.

4. J. Monod, *Chance and Necessity: An Essay on the Natural Philosophy of Modern Biology* (New York: Knopf, 1971).

5. It may be that understanding how God governs the creation is simply beyond us. Certainly it raises the specter of the predestination versus freewill debate. If God chooses to have the lots turn up a certain way, does God act to shift the outcome of their fall, perhaps invisibly though quantum indeterminacy? Or does God from eternity reset the initial state of the system, perhaps all the way back to the Big Bang, to get the desired result? But yet, the Scriptures state that God does so rule.

Chapter 8

1. P. Alberch and M. J. Blanco, "Evolutionary Patterns in Ontogenetic Transformation: From Laws to Regularities," *International Journal of Developmental Biology* 40, no. 4 (August 1996): 845–58.

2. J. Endler, *Natural Selection in the Wild* (Princeton, N.J.: Princeton University Press, 1986).

3. N. H. Barton and M. Turelli, "Evolutionary Quantitative Genetics: How Little Do We Know?" *Annual Review of Genetics* 23 (1989): 337–70.
4. The HOX genes are sets of genes that identify cellular location in animals. They come in sets arranged in order on the chromosome, and they are turned on in sequential order down the long axis of the animal. The sets of HOX genes from round worms, flies, and mice have the same homologous (matching) genes in sequence.
5. The reader who wants to pursue the role of genes in developing embryos needs to be prepared to do some deep wading. Modern embryology texts are essentially illustrated genetics texts. Developmental biology is being completely transformed about every five years.
6. C. Hodge, *What Is Darwinism?* (New York: Scribner, Armstrong, 1874).
7. D. Lamoureux, "Theological Insights from Charles Darwin," *Perspectives on Science and Christian Faith 56*, no. 1 (2004): 2–12.
8. B. B. Warfield, "Charles Darwin's Religious Life," *Princeton Theological Review* (1888): 569–601.

Chapter 9
1. A great deal of the information summarized here is from various articles in D. Otte and J. A. Endler, eds., *Speciation and Its Consequences* (Sunderland, Mass.: Sinauer, 1989), 28–59; also H. E. H. Paterson, "The Recognition Concept of Species," in *Species and Speciation*, ed. E. S. Vrba (Pretoria: Transvaal Museum Monograph, 1985), 21–29.
2. J. L. Feder, C. A. Chilcote, and G. L. Bush, "Genetic Differentiation between Sympatric Host Races of the Apple Maggot Fly, *Rhagoletis pomonella*," *Nature* 336 (1988): 61–64.
3. D. B. Wake, K. P. Yanev, and M. M. Frelow, "Sympatry and Hybridization in a 'Ring Species': The Plethodontid Salamander *Ensatina eschscholtzii*," in Otte and Endler, eds., *Speciation and Its Consequences*, 134–57.
4. For the genetic evidence for polar bear origins see G. F. Shields et al., "Phylogeography of Mitochondrial DNA Variation in Brown Bears and Polar Bears," *Molecular Phylogenetics and Evolution* 15, no. 2 (May 2000): 319–26; S. L. Talbot and G. F. Shields, "Phylogeography of Brown Bears (*Ursus arctos*) of Alaska and Paraphyly within the Ursidae," *Molecular Phylogenetics and Evolution 5*, no. 3 (June 1996): 477–94.
5. R. A. Fisher, *The Genetical Theory of Natural Selection* (Oxford: Clarendon, 1930).

Chapter 10
1. W. Frair, "Baraminology—Classification of Created Organisms," *Creation Research Society Quarterly* 37, no. 2 (September 2000): 82–91.

2. P. Wellnhofer, "Archaeopteryx," *Scientific American* (May 1990): 70–77.

3. K. Padian and L. M. Chiappe, "The Origin of Birds and Their Flight," *Scientific American* (February 1998): 38–47.

4. L. M. Chiappe and G. J. Dyke, "The Mesozoic Radiation of Birds," *Annual Review of Ecology and Systematics* 33 (2002): 91–124.

5. N. Eldredge, *Unfinished Synthesis* (Oxford: Oxford University Press, 1985); S. J. Gould, "Is a New and General Theory of Evolution Emerging?" *Paleobiology* 6, no. 1 (1980): 119–30; S. J. Gould, "The Paradox of the First Tier: An Agenda for Paleobiology," *Paleobiology* 11, no. 1 (1985): 2–12.

6. B. Michaux, "Morphological Variation of Species through Time," *Biological Journal of the Linnean Society* 38 (1989): 239–55.

7. B. J. MacFadden and R. C. Hulbert, "Explosive Speciation at the Base of the Adaptive Radiation of Miocene Grazing Horses," *Nature* 336 (1988): 466–68.

8. N. H. Barton and M. Turelli, "Evolutionary Quantitative Genetics: How Little Do We Know?" *Annual Review of Genetics* 23 (1989): 337–70.

9. D. K. Belyaev, "Destabilizing Selection as a Factor in Domestication," *The Journal of Heredity* 70 (1979): 301–8.

10. C. J. Allender et al., "Divergent Selection during Speciation of Lake Malawi Cichlid Fishes Inferred from Parallel Radiations in Nuptial Coloration," *Proceedings of the National Academy of Sciences of the United States of America* 100, no. 24 (2003): 14074–79.

11. R. E. Glor et al., "Phylogenetic Analysis of Ecological and Morphological Diversification in Hispaniolan Trunk-ground Anoles (*Anolis Cybotes* Group)," *Evolution* 57, no. 10 (October 2003): 2383–97; A. G. Stenson, R. S. Thorpe, and A. Malhotra, "Evolutionary Differentiation of Bimaculatus Group Anoles Based on Analyses of mtDNA and Microsatellite Data," *Molecular Phylogenetics and Evolution* 32, no. 1 (July 2004): 1–10.

12. MacFadden and Hulbert, "Explosive Speciation."

13. These notes cite two of the many of the review papers written in the last few years. D. M. Seaborg, "Evolutionary Feedback: A New Mechanism for Stasis and Punctuated Evolutionary Change Based on Integration of the Organism," *Journal of Theoretical Biology* 198, no. 1 (1999):1–26; E. H. Davidson, D. R. McClay, and L. Hood, "Regulatory Gene Networks and the Properties of the Developmental Process," *Proceedings of the National Academy of Sciences of the United States of America* 100, no. 4 (2003): 1475–80.

14. J. Hermisson, T. F. Hansen, and G. P. Wagner, "Epistasis in Polygenic Traits and the Evolution of Genetic Architecture under Stabilizing Selection," *American Naturalist* 1, no. 5 (2003): 708–34.

15. S. Gilbert, J. M. Opitz, and R. A. Raff, "Review: Resynthesizing Evolutionary and Developmental Biology," *Developmental Biology* 173 (1996): 357–72; R. L. Carroll, "Towards a New Evolutionary Synthesis," *Trends in Ecology and Evolution* 15, no. 1 (2000): 27–32.

16. R. L. Carroll, *Vertebrate Paleontology and Evolution* (New York: Freeman, 1988).

17. J. Gatesy et al., "Stability of Cladistic Relationships between Cetacea and Higher-level Artiodactyl Taxa," *Systematic Biology* 48, no. 1 (March 1999): 6–20; M. Nikaido, A. P. Rooney, and N. Okada, "Phylogenetic Relationships among Cetartiodactyls Based on Insertions of Short and Long Interspersed Elements: Hippopotamuses Are the Closest Extant Relatives of Whales," *Proceedings of the National Academy of Sciences of the United States of America* 96, no. 18 (1999): 10261–66.

18. J. G. Thewissen and E. M. Williams, "The Early Radiations of Cetacea (Mammalia): Evolutionary Pattern and Developmental Correlations," *Annual Review of Ecology and Systematics* 33 (2002): 73–90.

19. J. G. Thewissen et al., "Skeletons of Terrestrial Cetaceans and the Relationship of Whales to Artiodactyls," *Nature* 413 (2001): 277–81; P. D. Gingerich et al., "Origin of Whales from Early Artiodactyls: Hands and Feet of Eocene Protocetidae from Pakistan," *Science* 293, no. 5538 (2001): 2239–42.

Chapter 11

1. The literature on these techniques is massive. For recent reviews of the data see M. J. Sanderson and H. B. Shaffer, "Troubleshooting Molecular Phylogenetic Analyses," *Annual Review of Ecology and Systematics* 33 (2002): 49–72; B. S. Arbogast et al., "Estimating Divergence Times from Molecular Data on Phylogenetic and Population Genetic Timescales," *Annual Review of Ecology and Systematics* 33 (2002): 707–40; A. B. Smith and K. J. Peterson, "Dating the Time of Origin of Major Clades," *Annual Review of Earth and Planetary Science* 30 (2002): 65–88.

2. C. E. Nelson, B. M. Hersh, and S. B. Carroll, "The Regulatory Content of Intergenic DNA Shapes Genome Architecture," *Genomic Biology* 5, no. 4 (2004): R25; R. S. Mann and S. B. Carroll, "Molecular Mechanisms of Selector Gene Function and Evolution," *Current Opinion in Genetics and Development* 12, no. 5 (2002): 592–600; M. Ronshaugen, N. McGinnis, and W. McGinnis, "Hox Protein Mutation and Macroevolution of the Insect Body Plan," *Nature* 415 (2002): 914–17.

3. C. de Muizon, R. L. Cifelli, and R. C. Paz, "The Origin of the Dog-like Borhyaenoid Marsupials of South America," *Nature* 389 (1997): 486–89; J. A. Graves and M. Westerman, "Marsupial Genetics and Genomics," *Trends in Genetics* 18, no. 10 (2002): 517–21; see also the discussion of parallel evolution in Simon Conway Morris, *Life's Solution: Inevitable Humans in a Lonely Universe* (Cambridge: Cambridge University Press, 2003).

4. F. Crick, *Life Itself: Its Origin and Nature* (New York: Simon & Schuster,

1981); F. Hoyle and C. Wickramasinghe, *Evolution from Space: A Theory of Cosmic Creationism* (New York: Simon & Schuster, 1981).

Chapter 12

1. As previously noted, Simon Conway Morris, *The Crucible of Creation: The Burgess Shale and the Rise of the Animals* (Oxford: Oxford University Press, 1998). Another interesting presentation is Steven Gould, *Wonderful Life: The Burgess Shale and the Nature of History* (New York: Norton, 1989). See also J. Valentine, "Prelude to the Cambrian Explosion," *Annual Review of Earth and Planetary Science* 30 (2002): 285–306; D. Campbell and K. B. Miller, "The 'Cambrian Explosion': A Challenge to Evolutionary Theory?" in *Perspectives on an Evolving Creation* (Grand Rapids: Eerdmans, 2003).

2. R. L. Carroll, "Towards a New Evolutionary Synthesis," *Trends in Ecology and Evolution* 15, no. 1 (2000), p. 28.

3. A. Adoutte et al., "The New Animal Phylogeny: Reliability and Implications," *Proceedings of the National Academy of Sciences of the United States of America* 97, no. 9 (2000): 4453–56.

4. J. Y. Chen et al., "Small Bilaterian Fossils from 40 to 55 Million Years before the Cambrian," *Science* 305, no. 5681 (2004): 218–22.

5. K. J. Peterson et al., "Estimating Metazoan Divergence Times with a Molecular Clock," *Proceedings of the National Academy of Sciences of the United States of America* 101, no. 17 (2004): 6536–41.

6. A. H. Knoll and S. B. Carroll, "Early Animal Evolution: Emerging Views from Comparative Biology and Geology," *Science* 284, no. 5423 (1999): 2129–37.

7. E. H. Davidson, D. R. McClay, and L. Hood, "Regulatory Gene Networks and the Properties of the Developmental Process," *Proceedings of the National Academy of Sciences of the United States of America* 100, no. 4 (2003): 1475–80; K. J. Peterson and E. H. Davidson, "Regulatory Evolution and the Origin of the Bilaterians," *Proceedings of the National Academy of Sciences of the United States of America* 97, no. 9 (2000): 4430–33.

8. D. L. Stern, "Evolutionary Developmental Biology and the Problem of Variation," *Evolution International Journal of Organic Evolution* 54, no. 4 (2000): 1079–91.

Chapter 13

1. Quoted from lesson materials produced in 1988 by the Word of Life Institute of Schroon Lake, New York.

2. The evidence for the Australopithecines has been widely documented in sources ranging from *Newsweek* to *Paleobiology*. Notable are the series of books produced by paleoanthropologists Donald Johanson and Richard Leakey. See, for

instance, D. Johanson and M. Edey, *Lucy: The Beginnings of Mankind* (New York: Simon & Schuster, 1981); R. E. Leakey and R. Lewin, *Origins: What New Discoveries Reveal about the Emergence of Our Species and Its Possible Future* (New York: Dutton, 1977). For more recent discussions see I. Tattersall, "Once We Were Not Alone," *Scientific American* (January 2000): 56–62; H. M. McHenrey and K. Coffing, "*Australopithecus* to *Homo:* Transformations in Body and Mind," *Annual Review of Anthropology* 29 (2000): 125–46.

3. For further reading about these early *Homo* finds see J. Hurd, "Hominids in the Garden?" in *Perspectives on an Evolving Creation*, ed. K. B. Miller (Grand Rapids: Eerdmans, 2003); A. Walker and P. Shipman, *The Wisdom of the Bones: In Search of Human Origins* (New York: Knopf, 1997); C. C. Swisher, G. H. Curtis, and R. Lewin, *Java Man: How Two Geologists' Dramatic Discoveries Changed Our Understanding of the Evolutionary Path to Modern Humans* (New York: Scribner, 2000).

4. R. Foley, "The Context of Human Genetic Evolution," *Genome Research* 8, no. 4 (1998): 339–47; D. E. Lieberman, B. M. McBratney, and G. Krovitz, "The Evolution and Development of Cranial Form in *Homo sapiens*," *Proceedings of the National Academy of Sciences of the United States of America* 99, no. 3 (2002): 1134–39; R. Foley, "The Ecological Conditions of Speciation: A Comparative Approach to the Origins of Anatomically Modern Humans," in *The Human Revolution: Behavioural and Biological Perspectives on the Origins of Modern Humans,* ed. P. Mellars and C. Stringer (Edinburgh: Edinburgh University Press, 1989), 298–320.

5. H. J. L. Valladas et al., "Thermoluminescence Dating of Mousterian 'Proto-Cro-Magnon' Remains from Israel and the Origin of Modern Man," *Nature* 331 (1988): 614–16; C. Stringer and R. McKie, *African Exodus: The Origins of Modern Humanity* (New York: Holt, 1997).

6. J. D. Clark et al., "Stratigraphic, Chronological and Behavioural Contexts of Pleistocene *Homo sapiens* from Middle Awash, Ethiopia," *Nature* 423 (2003): 747–52; T. D. White et al., "Pleistocene *Homo sapiens* from Middle Awash, Ethiopia," *Nature* 423 (2003): 742–47.

7. C. Stringer and C. Gamble, *In Search of the Neanderthals: Solving the Puzzle of Human Origins* (New York: Thames and Hudson, 1993).

8. G. Clark, "Image of God," in *Baker's Dictionary of Christian Ethics,* ed. C. F. H. Henry (Grand Rapids: Baker, 1973), 313.

9. G. C. Berkouwer, *Man: The Image of God,* trans. D. W. Jellema (Grand Rapids: Eerdmans, 1962), 179.

10. R. Anderson, *On Being Human: Essays in Theological Anthropology* (Grand Rapids: Eerdmans, 1982).

11. R. Foley, "The Context of Human Genetic Evolution," *Genomic Research* 4 (1998): 339–47.

12. P. Mellars, "Technological Changes at the Middle-Upper Palaeolithic

Transition: Economic, Social, and Cognitive Perspectives," in Mellars and Stringer, *The Human Revolution*, 338–65.

13. C. S. Henshilwood et al., "Emergence of Modern Human Behavior: Middle Stone Age Engravings from South Africa," *Science* 295, no. 5558 (2002): 1278–80; C. S. Henshilwood et al., "An Early Bone Tool Industry from the Middle Stone Age at Blombos Cave, South Africa: Implications for the Origins of Modern Human Behaviour, Symbolism and Language," *Journal of Human Evolution* 41, no. 6 (2001): 631–78; W. A. Niewoehner, "Behavioral Inferences from the Skhul/Qafzeh Early Modern Human Hand Remains," *Proceedings of the National Academy of Sciences of the United States of America* 98, no. 6 (2001): 2979–84.

14. Clark et al., "Stratigraphic, Chronological and Behavioural Contexts"; White et al., "Pleistocene *Homo sapiens* from Middle Awash."

15. A.-H. Salem et al., "Alu Elements and Hominid Phylogenetics," *Proceedings of the National Academy of Sciences of the United States of America* 100, no. 22 (2003): 12787–91.

16. G. Finlay, "*Homo divinus:* The Ape That Bears God's Image," *Science and Christian Belief* 15, no. 1 (2003): 17–40.

17. M. Ruvolo, "Genetic Diversity in Hominoid Primates," *Annual Review of Anthropology* 26 (1997): 515–40; P. Gagneux et al., "Mitochondrial Sequences Show Diverse Evolutionary Histories of African Hominoids," *Proceedings of the National Academy of Sciences of the United States of America* 96, no. 9 (1999): 5077–82; L. B. Jorde et al., "The Distribution of Human Genetic Diversity: A Comparison of Mitochondrial Autosomal and Y Chromosome Data," *American Journal of Human Genetics* 66, no. 3 (2000): 979–88.

18. R. L. Cann, M. Stoneking, and A. C. Wilson, "Mitochondrial DNA and Human Evolution," *Nature* 32 (1987): 531–36.

19. The genetic evidence for human prehistory has been worked out though hundreds of research reports. A basic introduction can be gotten from B. Sykes, *The Seven Daughters of Eve: The Science That Reveals Our Genetic Ancestry* (New York: Norton, 2001); S. Olson, *Mapping Human History: Genes, Race, and Our Common Origins* (Boston: Houghton Mifflin, 2002).

20. J. W. Ijdo et al., "FRA2B Is Distinct from Inverted Telomere Repeat Arrays at 2q13," *Genomics* 12, no. 4 (1992): 833–35; J. W. Ijdo et al., "Origin of Human Chromosome 2: An Ancestral Telomere-Telomere Fusion," *Proceedings of the National Academy of Sciences of the United States of America* 88, no. 20 (1991): 9051–55; Finlay, "*Homo divinus*"; D. Wilcox, "Establishing Adam: Recent Evidences for a Late-Date Adam (AMH @ 100,000 BP)," *Perspectives on Science and Christian Faith* 56, no. 1 (2004), pp. 49–54.

21. D. E. Lieberman, B. M. McBratney, G. and Krovitz, "The Evolution and Development of Cranial Form in *Homo sapiens*," *Proceedings of the National*

Academy of Sciences of the United States of America 99, no. 3 (2002): 1134–39; Wilcox, "Establishing Adam."

22. Sykes, *The Seven Daughters of Eve*; Olson, *Mapping Human History.*

23. R. W. Schmitz et al., "The Neandertal Type Site Revisited: Interdisciplinary Investigations of Skeletal Remains from the Neander Valley, Germany," *Proceedings of the National Academy of Sciences of the United States of America* 99, no. 20 (2002): 13342–47; A. Cooper, A. J. Drummond, and E. Willerslev, "Ancient DNA: Would the Real Neandertal Please Stand Up?" *Current Biology* 14, no. 11 (2004): R431–33; D. Caramelli et al., "Evidence for a Genetic Discontinuity between Neandertals and 24,000-Year-Old Anatomically Modern Europeans," *Proceedings of the National Academy of Sciences of the United States of America* 100, no. 11 (2003): 6593–97.

24. T. F. Bergstrom et al., "Recent Origin of HLA-DRB1 Alleles and Implications for Human Evolution," *Nature Genetics* 18, no. 3 (1998): 237–42.

Chapter 14

1. C. Hodge, *Systematic Theology, Volume 1* (New York: Scribner, Armstrong, 1872): 571, 574.

2. Augustine, *The Literal Meaning of Genesis, Volume 1* trans. John Hammond Taylor (New York: Newman, 1982): 42–43.

3. E. J. Larson and L. Witham, "Scientists and Religion in America," *Scientific American* (September 1999): 88–93.

4. M. Noll, *The Scandal of the Evangelical Mind* (Grand Rapids: Eerdmans, 1994), 198.

5. For further discussion by evangelical theologians of what it means to read the Scripture as it was meant to be read in such faith-science issues, I recommend B. K. Waltke, "The Literary Genre of Genesis, Chapter One," *Crux* 27 (1991): 2–10; H. Blocher, *In the Beginning: The Opening Chapters of Genesis* (Downers Grove, Ill.: InterVarsity Press, 1984).